Highway under the Hudson

Robert W. Jackson

Highway under the Hudson

A History of the Holland Tunnel

New York University Press • *New York and London*

NEW YORK UNIVERSITY PRESS
New York and London
www.nyupress.org

References to Internet websites (URLs) were accurate at the time of writing.
Neither the author nor New York University Press is responsible for URLs
that may have expired or changed since the manuscript was prepared.

Library of Congress Cataloging-in-Publication Data
Jackson, Robert W. (Robert Wendell), 1950–
Highway under the Hudson : a history of the Holland Tunnel /
Robert W. Jackson.
p.   cm.
Includes bibliographical references and index.
ISBN 978-0-8147-4299-0 (cloth : alk. paper) — ISBN 978-0-8147-4504-5
(ebook) — ISBN 978-0-8147-4503-8 (ebook)
1. Holland Tunnel (New York, N.Y.)—History. 2. Tunnels—New York
(State)—New York—Design and construction—History—20th century.
3. City planning—New York (State)—New York—History—20th century.
4. Urban transportation—New York (State)—New York—History—20th
century. 5. New York (N.Y.)—History—1898–1951. 6. New York (N.Y.)—
Economic conditions—20th century. 7. New York (N.Y.)—Politics and
government—1898–1951.   I. Title.
TA820.N44J33 2011
388.4'11—dc23   2011028188

New York University Press books are printed on acid-free paper,
and their binding materials are chosen for strength and durability.
We strive to use environmentally responsible suppliers and materials
to the greatest extent possible in publishing our books.

Manufactured in the United States of America

10 9 8 7 6 5 4 3 2 1

Contents

Illustrations appear in two groups, following pages 116 and 164.

Introduction

THE HOLLAND TUNNEL is one of two highway tunnels under the Hudson River between northern New Jersey and Manhattan Island in New York City (the Lincoln Tunnel is the other), and it was the first constructed. The one tunnel is actually composed of two connected parallel tubes, which are about 35 feet apart under the river but divert even farther as they rise to street level so that the entrances and exits at each end are separated by about two blocks to reduce traffic congestion. The north tube, with two lanes of westbound traffic, is 8,558 feet long from portal to portal. The south tube, with two lanes of eastbound traffic, is 8,371 feet long from portal to portal. Each tube is 29.5 feet in external diameter. The two tubes are commonly referred to as a singular facility, as they often will be in this history, in recognition of their physical connection and unified function.

When the Holland Tunnel opened in 1927, it was the longest and largest vehicular tunnel in the world, and the first with a ventilation system specifically designed to accommodate motor-vehicular traffic. Created in response to extensive research which determined the amount of carbon monoxide produced by motor vehicles, the safe limits of human exposure to this deadly gas, and the most cost-efficient method of providing fresh air and exhausting polluted air, the Holland Tunnel's ventilation system became the model for, or informed the design of, virtually every vehicular tunnel built thereafter.

The Holland Tunnel's unprecedented length, size, and ventilation system are enough to make it historically significant as a great achievement in civil and mechanical engineering. The role it has played from the day it opened to the present as a vital link in the transportation system of the New York metropolitan area also makes it a historical landmark of continuing importance. It proved the viability of vehicular tunnels as alternatives to the ferry-and-barge-based transportation systems of the Port of New York. As the automobile came of age and automobile registrations experienced explosive growth in New York and New Jersey, the tunnel met the need for an alternative to railroad-based transportation. In fact, the tunnel's very existence created a new need for "superhighways" in New York and New Jersey to handle the traffic that it generated—highways that were the first of their kind.

From 1904, when the first New York City subway line opened, to the early 1920s, most people relied on the subway, streetcars, and their own feet to get about in Manhattan. In the first decade of the twentieth century, the upper class used horse-drawn carriages or, like their upper-middle-class counterparts, occasionally used horse-drawn cabs. Horses also provided the motive power for commercial wagons, by which almost all freight moved throughout the city. The trade-off for dependence on the horse was the near-constant sight and smell of horse manure and urine on city streets, and the not-infrequent dead horse lying temporarily untended on the cobblestones. Gradually, however, the problematic presence of horses was supplanted by the also unpleasant sight, smell, and noise of motor vehicles.

In 1901, New York became the first state to require registration of motor vehicles, with New Jersey following in 1903. But it was not until 1921 that all states required annual registration of new vehicles, and reporting practices varied considerably among states. It is difficult, therefore, to know exactly how many motor vehicles were registered before 1921, much less how many existed (not all owners in the early years followed registration laws), but statistics available from the United States Federal Highway Administration (FHWA) provide a reasonable overview of trends. According to FHWA data, during the first two decades of the twentieth century, New York consistently led all states in the number of motor-vehicle registrations, and New Jersey was near the top of the list.[1]

In 1905, the year that the first gasoline-powered buses in the United States began operating on Fifth Avenue in New York, there were 77,800 motor-vehicle registrations in the United States.[2] Approximately 98 percent were private automobiles, and the rest were trucks. In New York State, there were 9,230 total motor-vehicle registrations in that year (including 160 trucks), and in New Jersey there were 3,640 (70 trucks).[3]

In January 1906, Henry Ford introduced his Model N at the Association of Licensed Automobile Manufacturers Automobile Show in Madison Square Garden, one of fifty-six different models displayed.[4] This car marked a watershed advance in automobile development, due to its simplicity of design, practicality, and above all, low cost. The estimated price announced at the show was $300, quite a contrast to the majority of the cars displayed, priced in the thousands of dollars. Although the price soon doubled to $600, the Model N was still a relatively inexpensive car.[5]

By the end of 1907, gasoline-powered taxis virtually replaced horse-drawn cabs in the Battery area of Lower Manhattan. There were 143,200 total new motor-vehicle registrations in the United States that year, including 2,900 trucks. In New York State, there were 11,750 registrations, including 240 trucks, and in New Jersey, there were 4,550 registrations, including 100 trucks. In 1907, annual registration figures reflected only new registrations and not the total number of vehicles already registered and in use.

According to the best estimates compiled by the Automobile Directories Company of New York City, there were approximately 230,000 motor vehicles registered and in use in the United States, as of September 1, 1917. Counting cumulative registrations from 1901 to 1907, New York led all states, with approximately 47,000 total automobile registrations, and New Jersey was second, with approximately 32,000.[6] (From this point forward, only the number of vehicles newly registered each year will be reported.)

A year later, Henry Ford introduced the Model T passenger vehicle, which was soon to become nearly ubiquitous throughout the country. It was the archetype of the mass-produced, affordable, easy-to-maintain, middle-class car.[7] As sales increased and production efficiency improved during the model's production from 1908 to 1927 (the year the Holland Tunnel opened), the company actually lowered the sales price from $850 to below $300.[8]

In January 1912, self-starters were introduced in a number of passenger vehicles during Phase I of the Twelfth National Automobile Show in Manhattan's Madison Square Garden, and a large number of models had electric lights, replacing the troublesome kerosene lamps of earlier models. These improvements marked an important advance in the practicality and usefulness of automobiles. The greatest leap forward in design, however, appeared during Phase II of the show, which brought in large crowds to inspect the wide variety of trucks then being manufactured. In previous years, commercial vehicles had been relegated to the basement or secondary viewing areas of the show, while "pleasure cars," as passenger automobiles were still called, dominated the prime floor space. Now trucks received greater attention, and two weeks were devoted solely to commercial vehicles after the passenger-vehicle phase of the show ended. Harry S. Houpt, one of the show organizers, said, "Businessmen have seen with their own eyes that the truck has passed out of the experimental stage, and is a vital factor in their transportation systems."[9]

Among the interested spectators on January 11 was William F. "Buffalo Bill" Cody, the frontier icon whose Wild West Show had so often filled the same arena. The *New York Times* reporter who interviewed the great show-man found that Cody always "had a great love of horse flesh, but he admits that the motor propelled vehicle has an advantage over the equine drawn wagon nowadays."[10]

The following month, the Motor Vehicle Dealers' Association held its second annual, week-long automobile show in the Twenty-third Regiment Armory in Brooklyn. This exhibition featured an appearance by Joan New-ton Cuneo, "America's premier woman motorist," and offered "special at-tractions" for women.[11]

Less than a month later, on April 13, 1912, the Motor Truck Club spon-sored its second annual parade in Manhattan, showcasing five hundred trucks representing fifty-three different manufacturers. The line of trucks stretched nearly two miles along the parade route, which ran from the Bat-tery to 125th Street. According to the *New York Times*, the parade was "a revelation to the public of the giant strides which are taking place in the evolution of power transportation vehicles."[12] Many of the manufacturers represented in the parade were local, such as the Hewitt Motor Company, a New York City–based builder of highly engineered motor trucks.

Hewitt was one of many manufacturers whose products contributed to a total U.S. vehicular registration of 944,000 in 1912 (including 42,404 trucks). New York had more registrations than any other state, with 107,260 (including 4,720 trucks), and New Jersey had 35,410 (including 1,540 trucks).

By the end of 1916, the year Congress passed the Federal Aid Road Act, thus greatly facilitating the construction of highways across the United States, there were more than 3.6 million motor-vehicle registrations in the country. This amounted to one vehicle for every three persons living in the United States, an increase of more than 283 percent from 1912, with a remarkable 489 percent growth in truck registration. New York's vehicle registrations nearly doubled from 1912 to 1916, with 314,222 registrations in 1916. But even this rate of growth pales in comparison to the 644 percent rise in truck registrations from 1912 to 1916. New Jersey's registrations grew to 109,414 (a 209 percent increase from 1912), while its percentage increase in truck registrations outpaced even that of New York, at approximately 816 percent.

In 1917, the United States' entry into World War I greatly increased the demand for trucks due to the need to transport war material in this country and abroad. Ford introduced the Model T one-ton truck chassis that year, its first chassis built specifically for trucks. New York City discontinued its last horse-drawn streetcar line in 1917, but horse-drawn wagons were still widely used by the trucking industry.

There were 5,118,525 motor-vehicle registrations in the United States in 1917, with 391,057 of those being trucks. New York again led the nation in number of motor vehicles registered in 1917, with 406,016 (55,402 trucks), while New Jersey also showed impressive growth in registrations, with 141,918 (22,300 trucks).[13]

In 1920, the year that construction began on the Holland Tunnel, motor-vehicle registrations had increased almost 81 percent over 1917 to more than 9 million, and truck registrations had increased by 183 percent to 1.1 million. New York had 676,205 total registrations that year (an increase of almost 67 percent), including 125,401 trucks (a 126 percent increase), while New Jersey had 227,737 total registrations (an increase of 60 percent), including 50,400 trucks (an increase of 126 percent).

Before the Holland Tunnel opened, there was no road or highway to carry the ever-increasing vehicular traffic between the nation's largest city and New Jersey. Ferries owned and operated by railroad corporations carried almost all passenger automobiles, trucks, and horse-drawn vehicles that crossed the Hudson River into or out of Manhattan. Ferry service was often less than serviceable, however, and was frequently delayed or stopped altogether due to fog or ice in the river or to labor strikes, which were common in the turbulent early decades of the twentieth century.

For daily commuters or railroad passengers, there were options available. The Hudson and Manhattan Railroad began transporting people under the river to Manhattan subway stations at midtown in 1908 and to a downtown station in 1909. Farther uptown, the Pennsylvania Railroad completed a passenger-train tunnel under the river into Pennsylvania Terminal (Penn Station) in 1910. But there were no options for motor vehicles or horse-drawn wagons. Even on days of normal operation, they might have to stand idle for hours in long lines stretching miles through the streets of Jersey City, waiting to board one of the overtaxed ferryboats. In Lower Manhattan, the situation was similar for vehicles headed west to New Jersey, with near-constant traffic gridlock.

Despite rapid growth in the motor-trucking industry during the early years of the twentieth century, the majority of bulk freight from New Jersey destined for delivery to the boroughs of New York City, including almost all the city's food and fuel, was transported across the river or through the harbor by small, self-propelled cargo boats, called "lighters"; in barges pushed by towboats; or in towboat-propelled "car floats" carrying railcars filled with freight. Lighters, barges, and car floats also carried freight to railroad-owned terminals in New York City for transshipment to other parts of the Northeast or to the docks along Lower Manhattan's west side, where virtually all the oceangoing ships using the Port of New York loaded their cargo. Although many of the watercraft making up this system, particularly the towboats, were independently owned and operated, the railroad corporations dominated the shipping business of the port.

Like the ferries, this system was vulnerable to the vagaries of weather and labor unrest. Its inadequacies created a problem not just for the greater New York area but for the entire country. In the years immediately preceding and following World War I, approximately 50 percent of the nation's foreign trade passed through the Port of New York.[14]

During the severe winter of 1917–1918, the Hudson River was so covered and clogged with ice that fuel and food could not be brought into the city in sufficient quantities to meet the needs of the population. The heating and hunger crisis that ensued, along with the backup of vital war materials destined for Europe in the Port of New York, exposed with dramatic clarity the existing railroad-controlled transportation system's fatal flaws. The need for an alternative was so apparent that opposition to a highway across the Hudson River began to break apart in 1918 as rapidly as did the ice jams in the river as warmer weather returned. But would that highway be brought across the river by a bridge going over or a tunnel going under?

As early as 1906, state commissions existed in New Jersey and New York charged with the responsibility of determining how best to answer that question. In the first six years of their existence, the commissions primarily concerned themselves with developing plans for a rail, wagon, and pedestrian bridge. In this they followed the precedent established by private-venture planners, who, as early as 1868, had sought charters at the state or federal level for construction of a bridge, a structure that initially seemed easier (and less costly) to construct than a tunnel.

In 1913, the state commissions resolved to build a tunnel first. Engineering studies had revealed that a tunnel was the most feasible and cost-effective option for a link from northern New Jersey to Lower Manhattan, the commercial heart of New York City. A bridge located farther upstream, or additional tunnels, could come later. Questions remained, however, regarding the best type of tunnel and the optimal method of construction.

By 1913, tunnel engineers had already accumulated considerable experience in building rail and utility tunnels under rivers in two ways. They could dig through the bed of the river, using compressed air chambers to keep the water out until the tunnel was finished, or they could assemble the tunnel sections on shore and then sink them into a preexcavated trench. Each method had advantages and disadvantages. There were also options for the material used to construct the tunnels. Cast iron was the most tried and true, but steel and concrete had their advocates, as well.

What tunnel engineers had not yet figured out was the best method of ventilating a vehicular tunnel. There were a few, relatively short tunnels used by motor vehicles in Europe, but these had been built for use by horse-drawn wagons. As vehicular traffic in them grew, it became increasingly evident that the European tunnels were limited in the amount of traffic they could accommodate without risking human health. Some experts were certain that a vehicular tunnel long enough to reach from New Jersey to New York could not be properly ventilated by any method. Perhaps, they proposed, freight rail tunnels for trains pulled by electric locomotives or tunnels for unmanned carts pulled by cables would be a better and safer option. Even so, the state commissions were committed to constructing a vehicular tunnel. But should such a tunnel be a single-tube design or a multiple tube? And how many lanes of traffic should it accommodate?

These technical questions were largely the purview of the state commissions' staff engineers and consulting engineers, who, in application of the Progressive-era ideal of objective problem-solving, sought to overcome the physical barrier of the river through the application of scientific study. Years of research followed as they attempted to address these questions, while the commissioners worked to secure the legislative authorization necessary to fund construction. Before detailed design studies could begin, the state commissions had to overcome the political barriers that separated the two states and separated Democratic and Republican politicians within each state.

Not until 1920 were questions regarding type, method of construction, configuration, and funding source resolved and construction begun. Once started, the project took more years, cost more money, and sacrificed more lives than anyone could have imagined.

In *The Road and Car in American Life* (1971), John B. Rae cites road building, including "adjuncts such as bridges and tunnels," as "economic activity on a massive scale that has been created directly by the automobile and that has been paid for largely by the automobile as well. Apart from the economic effects, the motor vehicle has also been responsible for some spectacular achievements in civil engineering that would not have been undertaken without the demand created by automobile travel."[15] Rae's partial list of U.S. projects includes San Francisco's Golden Gate and Bay Bridges, the Lincoln and Holland Tunnels, the bridges between New Jersey and New York, and the Mackinac Bridge. What he does not note is that the Holland Tunnel was the first of these to be built. It was, therefore, a seminal work in the history of urban transportation.

The history that follows tells how a vitally important link in the regional transportation system of the New York metropolitan area came to be, as it describes an unprecedented achievement in the fields of civil and mechanical engineering. It is also a tale of great human drama, with heroes and villains, that illustrates how great things are accomplished, and at what price.

1

The Impetus

The impetus which was given to the building of the great Holland Tunnel was undoubtedly the most notable constructive achievement of my first term as Governor.

—Walter Evans Edge, *A Jerseyman's Journal:*
Fifty Years of American Business and Politics, 1948

THE TWO MEN met, at the governor's request, in the dining hall of the Army and Navy Club on the corner of Connecticut Avenue and I Street NW, in Washington, D.C. It was the middle of January 1917. One of the men, Walter E. Edge, had just been elected chief executive of the state of New Jersey. During his campaign and in his inaugural address, Edge committed his administration to modernizing New Jersey's transportation infrastructure. His three-part plan included a vehicular bridge across the Delaware River between Camden, New Jersey, and Philadelphia, Pennsylvania; a vehicular tunnel under the Hudson River between Jersey City, New Jersey, and New York City; and overhaul of the methodology for planning and building state highways.

The first two steps would relieve the ever-growing congestion of automobile and truck traffic between New Jersey and the two great metropolitan areas adjoining the state, Philadelphia to the west and New York to the east. The existing river-ferry systems were overburdened and incapable of meeting fully the increasing needs of commuters and freight haulers.[1] The third step would improve the poor quality of state roads, which, under the existing system, were built by inefficient, politically motivated, and often corrupt county organizations.

Edge wanted to place highway construction in the hands of state-employed technocrats. The federal government, which had become an important source of funding for highway expansion, wanted the same thing. In 1916, Congress passed the Federal Aid Road Act, which set aside

$75 million to be spent over five years for construction of "any public road over which the United States mails now are or may hereafter be transported." Under this new law, the federal government would match any funds the states expended to improve roads that met certain federal standards. Any state that applied for federal aid, however, had to establish a state highway department, staffed by qualified engineers.

Edge's three-step plan would be difficult to accomplish all at once. Politicians in northern New Jersey opposed the Delaware River bridge project, which they thought primarily benefited the southern counties. The Hudson River tunnel was opposed by those in southern New Jersey who saw no reason to fund a facility primarily benefiting the populous northern counties of Hudson, Essex, Bergen, and Passaic. Reorganizing the highway program had broader support because it promised to benefit the entire state, and Edge believed that he knew how to pay for the initiative. He proposed a state tax that could, over five years, provide a pool of $15 million into which he could dip to pay for the new highway program. But in a state with a long history of governmental corruption on every level, there was considerable suspicion about how the money would be spent. He had to appoint someone to the position of state engineer who would be "beyond criticism," possibly a man from out of state but certainly someone with an impeccable record of professionalism and accomplishment. Edge believed that his dinner guest was that man: Major General George Washington Goethals, builder of the Panama Canal.

Born in Brooklyn on June 29, 1858, to Dutch parents of modest means, Goethals moved with his family to number 47 Avenue D in Manhattan when he was eleven. It was a quiet residential neighborhood, located about a block from the East River, populated by families whose men worked in shipbuilding and allied trades. Intelligent and eager to learn, he determined early on to succeed in the world by studying a profession, perhaps medicine or the law. He was only fourteen when he began nearly four years of self-supported study at the College of the City of New York, where he discovered a talent for mathematics. He left prior to graduation after winning an appointment to the United States Military Academy, where he became a cadet captain his first year and, eventually, class president. After graduating second in his class in 1880, he worked for many years as an engineer on various river and harbor improvement projects, also serving stateside as chief engineer of I Corps during the Spanish-American War. He returned to the academy to teach civil and military engineering from 1885 to 1889 and again from 1898 to 1900.[2]

In 1907, President Theodore Roosevelt appointed Goethals as chief engineer of the Panama Canal project, begun by the French when Goethals was still a cadet. His performance in that role, while not without criticism, was generally regarded as a great success and brought him widespread fame and recognition as a man who could get things done. After the canal opened in 1914, Goethals remained in Panama as governor of the Canal Zone until he sailed for New York in September 1916. He submitted his resignation as governor on January 17, 1917.

When he arrived in New York on October 2, reporters asked Goethals what he intended to do next. "Well," he replied, "I'm going to look around for a job."[3] Four days later, President Woodrow Wilson appointed him chairman of a board of inquiry charged with examining the effectiveness of the Adamson Act, passed in 1916 in order to force an eight-hour workday on the nation's railroads. In January 1917, Goethals joined a New York engineering firm, which thereupon changed its name to Goethals, Jamieson, Houston & Jay, with an office at 43 Exchange Place in Jersey City. Before he could begin work, however, he had to travel to Washington, D.C., in connection with his work on the Adamson board. Governor Edge, who had never met Goethals, telephoned him there at his hotel and invited him for dinner, without revealing the purpose of the meeting.

Goethals was about six feet tall, with intensely blue eyes and closely cropped gray hair, which he parted in the middle. Erect in bearing and still very fit at fifty-eight, he had the look of a commander of men, even in civilian clothes. Somewhat shy as a boy, he tended to appear intense and a bit stiff around those whom he did not know. He had two sons, to whom he expressed affection, but his relationship with his wife was "difficult," no doubt because of his slavish devotion to his work.[4]

In that work and in the work of others, Goethals believed in strict adherence to principles—his own. Goethals's biography in *Transactions of the American Society of Civil Engineers*, published after his death, notes, "he was a man of strong convictions, but open to argument and tolerant of difference from his own views, even of opposition to them." Once he made up his mind, however, and issued an order, he "expected loyal compliance, and did not welcome any attempt to revive the issue."[5]

At forty-three, Walter Edge was fifteen years younger than the retired general. Although an avid sportsman, he seemed softer in appearance than Goethals and was a bit jowly. Born in Philadelphia on November 20, 1873, Edge was two years old when his mother died. His father was employed by the Pennsylvania Railroad and remarried about the same time a new assignment to the company's West Jersey and Seashore division forced the family to relocate to Pleasantville, New Jersey. Unlike Goethals, Edge did not have much of a formal education, never going beyond eighth grade. But he, too, possessed a bright, active mind and a talent for self-education.

When Edge was fourteen, he began working for the *Atlantic Review*, Atlantic City's only newspaper. Later he went to work for a small advertising company, and at the age of seventeen he bought the agency when its owner died. Under his management, the business grew rapidly and provided Edge with a substantial income. He also continued in the newspaper business and was only twenty-one when he founded the *Atlantic City Daily Press*, which became the city's dominant newspaper. The advertising agency and the newspaper made Edge wealthy and allowed him to turn his attention to politics.

After moving to Atlantic City, Edge joined the local Republican Club. In 1894, at twenty years old and still three months shy of having the right to vote, Edge won election to the Republican Party city executive committee. Three years later, he began his involvement in state politics, serving as journal clerk of the New Jersey Senate until 1899 and as secretary of the senate from 1901 to 1904. Edge ran for election as a state senator in 1904 but lost. He ran again in 1909 and won, and he served two consecutive three-year terms, becoming senate president in 1915.

Edge decided to run for governor of New Jersey in 1916 as a "businessman with a business plan." His strongest opponent for the Republican nomination, however, was also a man with a strong business background, Austen Colgate of the Colgate Soap Company. A progressive from Essex County, Colgate could count on support from the densely populated northern counties containing the state's two largest cities, Newark (in Essex County) and Jersey City (in Hudson County). It was estimated that the counties of Bergen, Hudson, and Essex alone would account for more than a third of the Republican vote in the state. Yet Edge won by about three thousand votes, the narrowest margin in New Jersey history for a primary involving the election of a governor or U.S. senator.[6]

Jersey City political boss Frank Hague, a Democrat, instructed those whom he controlled to "cross over" and vote for Edge in the Republican primary, thus securing the victory.[7] Edge easily defeated the Democratic candidate, former Jersey City mayor H. Otto "The Dutchman" Wittpenn, in the general election. That was what Hague desired, fearing that election of the reform-mined Wittpenn might threaten Hague's plans for control of city, county, and even state politics.[8] With the election over, it was now time for Edge to deliver on his campaign promises—with Goethals's help.

As Edge and Goethals sat down to dine, and after sharing a few "formalities and pleasantries," Edge got straight to the point. "General," he said, "I am looking for the ablest engineer in the United States to fill the post of state engineer of New Jersey." He then outlined his transportation improvement program, emphasizing that highway construction would have first priority, with the Delaware River bridge and Hudson River tunnel projects to follow close behind. He also told the general that the position would require not only great technical expertise but also the ability to control the process free from political influence. It was exactly the type of job that suited Goethals's no-nonsense personality, and it would also pay extremely well. "General, my salary as Governor is $10,000 a year," Edge concluded. "If you will accept this position, I will guarantee you just double that amount: $20,000 a year."[9]

Although Edge made the offer unsure that he actually had the authority to pay for the position, he was more concerned that Goethals would turn him down flat. He need not have worried. Goethals seemed flattered by the offer, and after a thoughtful pause, he said, "Governor, I have practically completed my responsibilities in Panama. If the War Department will permit me, and I believe they will, and if other duties can be arranged, I am inclined to accept your offer."[10] The two men then got up from the table, shook hands, and parted. Within a few hours, Goethals telephoned Edge and accepted the post. All that remained was state senate confirmation.

With Goethals on board and a means of funding the highway program identified, Edge turned his attention to the problem of how to fund and win legislative support for the bridge and tunnel initiatives. Edge believed that he could "slip through the horns of the dilemma" by promising a quid pro quo whereby both projects would be funded. Legislators from northern counties would support the Delaware River bridge in order to get the tunnel, and southern legislators would support the Hudson River tunnel in order to get the bridge. He proposed legislation to fund the two projects on a pay-as-you-go plan via a state tax, similar to that which would provide for the highway program, to meet the advance expenses.[11] The costs would be reimbursed to the state over a period of years through tolls charged for use.

Edge thought that he had solved the problem of how to win the vote of state legislators, but he still faced a daunting political and financial barrier. The Hudson River tunnel project could not move forward without agreement from the New York legislature to fund New York's half of the tunnel's cost. Although Edge assured New Jersey legislators that he could get the New Yorkers to cooperate by showing them that the tunnel was in their best economic interest, the sell was going to be harder than he wanted to admit. Examination of the complicated history of freight and passenger movement across the Hudson River and throughout the harbor that both states shared helps illuminate the challenges that Edge faced.

Before the development of commercial aviation, there were three ways to transport people and freight across a navigable river or bay: by boat, bridge, or tunnel. Bridge and tunnel construction is very capital and labor intensive and requires considerable organizational and technical skill. Boat operation is much easier. The earliest attempts to profit by transferring people and freight across American waterways, therefore, were always by boat.

In 1661, the New Netherlands Council approved the first commercially operated ferry established by law between New Amsterdam and New Jersey. It ran three days a week from a point near the end of Whitehall Street in Manhattan to the end of Communipaw Avenue in Jersey City, near what later became the passenger terminal of the Central Railroad of New Jersey.[12]

The Dutch were thought to have selected the Communipaw location because it was where Henry Hudson first set foot on land while exploring the area. The Dutch, therefore, with their "remarkable tenacity . . . to do everything just as their predecessors did it," fixed that point as a landing place for all future generations.[13] The more logical explanation, however, is that anyone coming from Philadelphia wishing to take the shortest path across the river to the lower part of Manhattan Island, where the entire city of New York was then located, would naturally start from a point in that area. Hundreds of years later, Lower Manhattan was still the destination for the majority of traffic crossing from New Jersey.

In 1764, the second ferry operating to Jersey City (then called Paulus Hook) began service from the foot of Cortlandt Street in Manhattan. According to a 1909 account, the Paulus Hook and Communipaw ferries were "principally for the convenience of persons traveling between New York and Philadelphia. It is evident that Jersey City was considered in those days, as unfortunately it was for long afterward, not as a place to go, but only as a place to pass through when going somewhere else."[14] Of course, these ferries also served local traffic to and from Manhattan, as did the ferries with New Jersey termini further upriver. A ferry to Weehawken was established sometime before 1700, with a petition for further service approved in 1742. Until the Hoboken ferry began operation in 1775, the Weehawken ferry was the only service to and from New York for the farmers in the upper part of Bergen County.

These early ferries on the "North River," as the Hudson River between New Jersey and Manhattan was commonly known, were muscle-powered rowboats, whose passengers were occasionally forced to row, or wind-powered "periaugers," two-mast, flat-bottomed craft pointed at both ends, large enough to transport horses, carriages, and wagons. Usually, passengers and freight were not accommodated on the same boat. Distressingly slow, open to the weather, and easily carried off course by wind and the strong Hudson River tides, these early ferries hardly seemed worth the fare. Later, "horse boats"—propelled by teams of up to eight animals walking on a treadmill linked by gears to paddles—were used. This early technology did not mean a quicker or more pleasant voyage from one shore to the other, but the application of steam power in the early nineteenth century initially promised an improvement.[15]

In 1807, Robert Fulton began operation of the *North River Steamboat*, later called the *Clermont*, between New York City and Albany, New York. The venture was a commercial success, and within a few years Fulton, who had been granted the right to operate a ferry from Paulus Hook, ordered construction of two new boats for service between New York City and New Jersey. But a competitor, John Stevens, who also had been granted the right to operate a ferry, preceded Fulton in 1811 by placing the *Juliana* in service between Hoboken and Vesey Street in Manhattan. He found the craft unprofitable, however, and after little more than a year returned to the use of horse boats.

In 1812, Fulton began operating the *Jersey* between Jersey City and the foot of Cortlandt Street in New York, followed by the *York* in 1813. These paddle-wheel steamers, both double-hulled craft eighty feet long and thirty feet wide, were barely faster than the horse-powered boats, often taking an hour and a half to make a trip.[16]

It was not until the introduction of propeller-driven boats in the early 1890s that technological innovation substantially improved the quality of ferry service.[17] On January 11, 1893, New York Harbor was icebound completely for the first time in fifteen years, with conditions in the North River reported as the worst of all. Yet the Pennsylvania Railroad's propeller-driven ferryboats, recently placed in service, were able to push through the ice, while the paddle-wheelers of the other railroads never left their slips.[18]

The most significant change in ferry operation during the 1800s was the near-complete takeover by railroad corporations. The earliest railroads operating in northern New Jersey were content initially to establish terminals in places at some distance from New York, such as South Amboy or Elizabethport, and then to depend on steamboats to bridge the gap to Manhattan. Some boats carried only passengers, others only freight, while one Erie Railroad steamer carried only milk. Later, short-haul terminal railroads were created to transfer passengers, and eventually freight, from these more distant terminals to the Jersey shore terminals across from Lower Manhattan. Eventually, the long-haul railroads built their own terminal facilities on the Hudson River, starting with the Erie Railroad in 1862. Once these terminals were established, it was just a matter of time before business interests led the railroads to gain control of all ferry operations into Lower Manhattan from northern New Jersey.[19]

Beginning in 1886, the railroads developed an additional means of transferring freight across the river: car floats. The earliest of these wood barges had two tracks onto which up to three rail cars could be loaded by means of hinged bridges or ramps. Tugboats towed or pushed the floats across the river. At first, the railroads relied exclusively on independent operators, but later they came to own their own fleets of car floats, although a few private companies continued to provide service to private terminals. By 1917, the cross-river rail-car floats, some of which were made of steel and up to three hundred feet long, had three tracks and could carry from twelve to twenty-four rail cars.[20]

The great majority of freight transported in New York Harbor in 1917, however, was hauled by "lighters." Originally, these craft were small sail boats or, more commonly, flat-bottomed barges propelled by a tugboat, suitable for moving goods a short distance. Lighters were first owned and operated by independent companies, but over time each railroad developed its own fleet. Many private lighterage firms continued to serve steamship lines and share in a portion of the railroad-related business.[21]

The towboats that moved barges, lighters, and car floats were for the most part still operated by independent companies in 1917, with only the Pennsylvania Railroad and the Philadelphia and Reading Railroad operating their own towing fleets.

There was one notable exception to this process. Produce, milk, poultry, and certain other commodities were loaded on horse-drawn wagons or motor trucks at New Jersey terminals and brought across the river to Manhattan by ferry, along with pedestrians and privately owned automobiles.

In addition to the freight shipped across the Hudson River, the New York Central and Hudson River Railroad brought freight into Lower Manhattan via a line across Spuyten Duyvil Creek, which connects the Hudson and the Harlem Rivers. The rail line ran down the west side of Manhattan to a four-acre freight depot bounded by Hudson, Laight, Varick, and Beech Streets, on land purchased from Trinity Church in 1864. This site, formerly part of St. John's Park, was known as the St. John's Park Terminal. On Manhattan's west side, the railroad also operated a terminal at Thirty-Third Street, a terminal at Sixtieth Street, a milk station and other local service tracks at Manhattan Street (128th Street), and a small freight house on Dyckman Street (200th Street). Due to land development in the surrounding neighborhoods, however, the New York Central was unable to expand these facilities to keep up with demand and thus was forced, like its competitors, to rely on water transport of freight across the Hudson River.[22]

Due to the long waits to board the ferries, most trucking firms could not justify the cost of allowing motor trucks, which were expensive to own and operate, to sit idle in a long and slow-moving line. As late as 1917, the majority of "truck freight" in Manhattan and between Manhattan and New Jersey was still transported by horse-drawn wagons.[23]

Most of the railroad freight loaded in New Jersey onto car floats, and some of the freight that was lightered, was unloaded at so-called pier stations that, along with ferry terminals and docks for oceangoing ships, occupied virtually all the shoreline along the west side of Lower Manhattan. This New Jersey railroad freight was transferred from these pier stations to one of four "inland" terminals located between the piers and Eleventh Avenue, from Twenty-Third Street to Thirty-Ninth Street.[24]

The establishment of railroad freight piers and ferry docks along the west side of Lower Manhattan created a great deal of congestion in the streets of Lower Manhattan. "These streets," noted a *New York Times* article on March 9, 1919, "feed freight to the New York river fronts, and are congested with trucks going and coming from the forty-six piers used by the railroads. The trucks bear incoming and outgoing freight from lighters and cars. The crowding is so great that cars with outgoing freight frequently cannot be filled to capacity . . . and must be taken to the distributing freight yards on the Jersey side only partially loaded." And this was only part of the problem. As the *Times* also noted, "The ferries are so congested that long double lines of trucks [including horse-drawn and motor-vehicular "trucks"], extending for blocks, are constantly waiting at the approaches; this congestion causes delays and inconveniences, and is aggravated by storms, fog, ice, and heavy river traffic."[25]

Even though the railroad-operated system for transporting passengers and freight across the Hudson River had gained some acceptance by 1917, its inherent inefficiency had always been evident. From the late 1800s, therefore, there had been much talk in New Jersey and New York about bridging over or tunneling under the river as a means of replacing or, at least, augmenting the prevailing system.[26] It took an entrepreneur from the other side of the country, however, to transform talk into action.

One harsh winter day in March 1873, a group of California businessmen traveling to New York City were delayed overnight at the Jersey City ferry by a combination of fog and river ice. One of the travelers, Dewitt Clinton Haskins, had made a fortune in mining and railroad construction. During the delay, he suggested to his companions a permanent solution to the problem of bad weather: the formation of a company to operate trains under the Hudson River via a tunnel running from Jersey Avenue in Jersey City, under Fifteenth Street to the river, then under the bed of the river to the foot of Christopher Street in Manhattan and on to an eastern terminus at Broadway and Tenth Street. The railroad's primary cargo would be freight, with passengers a secondary consideration.[27] Before the year was out, Haskins incorporated the Hudson Tunnel Railroad Company, with charters in both New York and New Jersey.[28] Among his partners were U.S. Senator Roscoe Conkling of New York and Vermont financier Trenor W. Park.

In October 1874, Haskins began construction of a two-track tunnel, sinking a construction shaft at the foot of Fifteenth Street in Jersey City, but digging stopped after just a few weeks. The Delaware, Lackawanna and Western Railroad (Lackawanna), possibly fearing competition to its lighterage and ferry business, obtained an injunction, claiming infringement on property that it owned in the area. This was not the last time that a railroad corporation hindered construction of a tunnel terminating in Jersey City.

The claim was without merit, however, and after the tunnel company's rights were affirmed in court, construction resumed in September 1879. Plans then changed to include two tunnels with a single track in each. Haskins made considerable progress until July 1880, when twenty men drowned in the north tunnel. Although temporarily held up, the work did not stop completely, and the following year he began another construction shaft on the New York side at the foot of Morton Street.

Haskins's faith in the project began to waiver, however, after a series of engineering and financial difficulties caused periodic work stoppages. To make matters worse, both financial partners died, Park in 1882 and Conkling in 1888. An avid spiritualist, Haskins sought the help of a well-known

New York medium, who offered to put him in touch with his deceased partners. During sessions with the medium, Haskins received messages from the spirit world, written on the inside of two hinged, locked slates. A small bit of slate pencil was placed inside the tablets for the spirits' use. David Allyn Gordon, a friend of Haskins's who saw the slates, later recalled that Conkling's message was, "Be not dismayed; your enterprise will succeed," and "I abate no jot of my faith in your work." Park was similarly encouraging, writing, "Persevere with the tunnel. You will ultimately succeed," and "Don't be discouraged. You will win by and by."[29]

Despite the assurances of the dead, Haskins could not keep the enterprise going, and in 1899 the tunnels were sold under foreclosure. It was not until 1902 that the project began to move forward again, following formation of the New York and Jersey Railroad Company under the leadership of William Gibbs McAdoo, a prominent New York attorney. McAdoo intended that the tunnels would be used by passenger-only trains.

In 1903, McAdoo created a second organization, the Hudson and Manhattan Railroad Company, to construct two additional passenger-only "downtown" tunnels between Montgomery Street in Jersey City and Cortlandt Street in Manhattan. The more northern set of "uptown" tunnels (so called, even though they actually served what is now considered midtown) and the downtown tunnels would eventually be connected by a set of north-south tunnels on the Jersey City side.

The New Jersey and New York organizations were consolidated under the name "Hudson and Manhattan Railroad Company" on December 1, 1906. On February 25, 1908, the company began regular passenger service through the recently completed uptown tunnels, using smokeless electric trains. Service through the downtown tunnels followed on July 19, 1909. All the Hudson and Manhattan Railroad tunnels are now owned and operated by the Port Authority Trans-Hudson Corporation and are known as the PATH tunnels.

Some prognosticators believed that opening the Hudson and Manhattan tunnels (or, as they came to be called, to McAdoo's displeasure, the "McAdoo Tubes") meant that the ferryboats would eventually be relegated to the unprofitable transportation of wagons, trucks, and automobiles. They were mistaken.

The tunnels did initially divert a portion of the ferry passenger traffic, but explosive population growth in New Jersey and New York meant that the tunnels could never meet all passenger demand. As Captain John M. Emery, general marine superintendent of the Lackawanna Railroad, observed, "the tunnel cannot carry trucks, and a large part of our business consists of the transportation of trucks and automobiles." He added, "I do not believe the tunnel will cut into us very much even after the entire system is in full operation. Anyway, the number of commuters in New Jersey is constantly growing, so there will be enough business for us all."[30]

There was one other set of rail tunnels under the North River in 1917, and these also carried only passenger trains. On the same day that the first uptown McAdoo tube was "holed through," in March 1904, the Pennsylvania, New Jersey and New York Railroad, a subsidiary of the Pennsylvania Railroad, awarded a contract for construction of two rail tunnels between Weehawken, New Jersey, and Manhattan. A sister organization, the Pennsylvania, New York and Long Island Railroad, was also awarded a separate contract for construction of a set of tunnels under the East River to Long Island. These tunnels would serve a new Pennsylvania Railroad terminal (Penn Station) to be built at Thirty-Third Street in Manhattan. In 1907, the two corporations building the tunnels merged to form the Pennsylvania Tunnel and Terminal Railroad, and the railroad began regular service through the tunnels into Penn Station in 1910. With the opening of the station, and the phasing out of service to its old Jersey City passenger terminal, the Pennsylvania Railroad discontinued its Twenty-Third Street ferry, but vehicular and local pedestrian traffic kept its remaining ferries at Cortlandt Street and Desbrosses Street busy.

No other tunnels, and no bridges, had been built under or across the Hudson River between New Jersey and Manhattan by 1917. On the other side of the island, however, there were several links across the East River. The Brooklyn Bridge was the first (1883), followed by the Williamsburg Bridge (1903), the Queensboro Bridge (1909), and the Manhattan Bridge (1909). All of these carried a mixture of pedestrian, wagon, and motor-vehicular traffic to and from Brooklyn or Queens.[31]

And there were two subway tunnels in addition to the "heavy rail" tunnels of the Pennsylvania Railroad: the Battery-Joralemon Tunnel to Brooklyn (opened in 1908, it now carries the 4 and 5 trains) and the Steinway Tunnel to Queens (opened in 1915, it now carries the 7 train). There were also several more subway tunnels under construction: the Old Slip–Clark Street Tunnel (opened in 1919, it now carries the 2 and 3 trains); the Whitehall-Montague Tunnel (opened in 1920, it now carries the M and R trains); the East Sixtieth Street tunnel (opened in 1920, it now carries the N, Q, and R trains); and the Fourteenth Street Tunnel (opened in 1924, it now carries the L train).

Completion of vehicular bridges and rail tunnels across the East River encouraged economic expansion and population growth in the counties of Kings and Queens and also led to a general decline in the East River ferry system. Ferries remained, however, the only means of transferring vehicles across the Hudson River and, for some of the New Jersey railroads (the West Shore, Central of New Jersey, and, to a lesser extent, the Pennsylvania), of getting passengers to and from their terminals. In 1917, there were still thirteen ferry lines and fifty ferryboats serving Lower Manhattan across the Hudson, run by five railroads, with another two lines serving Upper Manhattan.[32]

Toward the end of the second decade of the twentieth century, a rapid rise in the number of vehicles needing to cross the Hudson River, along with general population growth in northern New Jersey and New York City, created a demand for service that the ferries could not meet. The barge-and-lighter system for transferring freight from New Jersey to New York City was also beginning to show signs of an inability to handle the needs of an expanding economy.

Since 1906, there were state commissions charged with the responsibility of determining how to improve transportation across the Hudson River by construction of either a bridge or a tunnel. By 1917, however, they had accomplished little, and as Steven Hart has observed in *The Last Three Miles* (2007), his insightful history of New Jersey's Pulaski Skyway, "the Hudson River remained a wall separating Manhattan from the road system gradually taking shape across the country."[33] But as great as the technological challenge of breaking through that wall via construction of a bridge or tunnel might have been, it was the difficulty of overcoming the problem of competing political and economic interests that had stymied, year after year, progress toward a solution.

2

Vexing Questions

I should say that the action of Governor Edge now points the way to conferences between the New Jersey and New York commissions that may lead to a settlement of the vexing but comparatively unimportant questions which have taken precedence over the larger aspects of this improvement.

—Frederick J. H. Kracke, Bridge Commissioner of
New York City, 1917

IN 1906, ELEVEN years before Walter Edge introduced his three-part transportation plan to New Jersey voters, the state legislature adopted a resolution creating the New Jersey Interstate Bridge Commission. The resolution cited the need for a more modern, rapid, and economical system of transit than that provided by the antiquated and sometimes dangerous system of ferryboats. It directed the three unpaid members of the commission to cooperate with a similar commission in New York State to study the feasibility of constructing a bridge over the Hudson River.[1]

In response to New Jersey's desire, in 1906 the New York legislature created the New York Interstate Bridge Commission. The commission, also composed of three unpaid members, was "to confer with the Governor and Legislature of the State of New Jersey for the purpose of developing a system of transit between the City of New York and the State of New Jersey."[2]

The two commissions met for the first time in New York City on June 14, 1906, and established a joint office at 115 Broadway. Both states enacted additional legislation for several years afterward, continuing the commissions, sometimes changing their names or the number of members, appropriating limited funds for commission needs, or extending their powers. These organizations accomplished little, however, other than to order borings of the Hudson River bed to determine the best location for bridge piers and abutments.

By 1912, many people had come to believe that a tunnel under the river might be a better option that a bridge over it. The New Jersey legislature therefore renamed the New Jersey Interstate Bridge and Tunnel Commission and authorized it to consider the possibility of a tunnel as well as a bridge. The New Jersey commission engaged the engineering firm of Jacobs and Davies, of London and New York, to submit a formal report the following year regarding a tunnel, while the New York State Bridge Commission hired the engineering firm of Boller, Hodge and Baird, of New York, to submit a report regarding a bridge.[3] When the New York legislature convened in 1913, it also authorized the newly renamed New York State Bridge and Tunnel Commission to consider the possibility of a tunnel in addition to a bridge.

Boller, Hodge and Baird submitted its official report to the New Jersey commission in February 1913 and to the New York commission in March, recommending a $42 million suspension bridge between Weehawken, New Jersey, and Fifty-Seventh Street (terminating at Ninth Avenue) in Manhattan. Experts already knew that in order to prevent obstruction of ship traffic in the Hudson River, as required by the War Department, the bottom of a bridge's road deck needed to be at least 150 feet above the water. There was a corresponding need to keep the approaches leading up to the main river span within a grade that vehicles could handle. If the land rose to meet the approaches, they could be kept short, thus saving the expense involved in purchasing land that would otherwise be required to accommodate long approaches with gradual slopes. But the land on both sides of the river below Fiftieth Street was too low for the construction of short approaches.

The proposed bridge would carry eight lines of rapid-transit trains, in addition to vehicular roadways. There would also be sidewalks for pedestrians. The terminus in Manhattan, the report concluded, would "afford communications with the Queensboro Bridge and the Queensboro Subway, thus affording direct communication between Long Island and New Jersey."[4] With a bridge site finally identified, the two commissions held a celebratory dinner on the night of March 19, 1913, at the Jersey City home of New Jersey commissioner J. Hollis Wells.[5]

In early April, Charles M. Jacobs and J. Vipond Davies submitted their official report laying out the argument for a tunnel and answering objections previously raised by bridge advocates. Jacobs and Davies had designed all six rail tunnels then under the Hudson: the two Pennsylvania Railroad and four Hudson and Manhattan Railroad tunnels. In addition,

Jacobs had served as chief engineer and Davies chief assistant engineer for construction of the first subaqueous (underwater) tunnel in the area, the East River Gas Company Tunnel between Manhattan and Brooklyn. Completed in 1894, the tunnel was large enough to walk through. The partners had also designed several subaqueous tunnels in England. These men knew how to build tunnels beneath a river. Their report noted that 19,660 vehicles a day crossed the Hudson River in each direction by ferry, but not more than 2,000 of those crossed north of Twenty-Third Street. Since a bridge was impracticable below Fifty-Seventh Street, they recommended construction of a tunnel from Canal Street in Manhattan to an extension of Twelfth Street in Jersey City, with an entrance and exit at about Henderson Street.[6]

The engineers selected the Canal Street terminus because all of the present freight-handling facilities on Manhattan congested in two districts along the river: one around Thirteenth Street and the other from Battery Park to West Tenth Street. This created constant difficulties. For example, the majority of food consumed in New York City was delivered to the very lower end of Manhattan. Lacking sufficient terminal facilities elsewhere, merchants as far away as the Bronx and Long Island were forced to haul their goods from these terminals.[7] Canal Street was a particularly wide avenue that led directly to the Manhattan Bridge and provided good access to the Brooklyn and Williamsburg Bridges, thus providing a direct route from the terminal districts to Long Island.

The engineers selected the Jersey City terminus because it would result in the shortest length for a tunnel with one end at Canal Street. With tunnels, construction costs rise exponentially with any increase in length; therefore, all other considerations being equal, the shorter the tunnel, the less expensive to construct. In addition, a Jersey City terminus would provide the best access to that area in New Jersey that had the greatest need for a tunnel into New York City.[8]

The tunnel should be a twin tube, the engineers proposed, with eastbound traffic in one tube and westbound traffic in the other, along a roadway seventeen feet wide. The tubes would have a capacity of five million vehicles per year, equaling approximately the entire vehicular traffic then crossing the Hudson River by ferries. The two tubes, with exterior sheaths made of steel, would be built by boring through the riverbed using the shield method. This method was first employed by Marc Isambard Brunel in construction of the world's first tunnel under a navigable river, the thirteen-hundred-foot-long Thames Tunnel in London, England. Built between 1825 and 1843, this subway tunnel was originally designed for use by horse-drawn carriages and wagons but was never used for that purpose.

The primitive shield method used by Brunel involved the construction of a three-story brick working chamber, with the shield located at the front in the direction of travel. The shield was subdivided by cross frames into thirty-six smaller working cells, which were open at the rear. At the front of the shield, movable boards were placed against the earth and kept in place by braces. Workers would remove one or two boards at a time, usually from the top down, and excavate the exposed earth in front of them. They would then place the boards back against the newly excavated surface and reapply the braces. When enough earth had been removed, the shield was shoved forward into the excavated space by screws, one at the top and one at the bottom, which, when turned, pushed against the finished brick walls behind and the boards in front. As the shield advanced, more bricks were added to the walls behind, forming a new surface against which the screws would push during the next shove. The tunnel thus advanced through the bed of the river in a series of shoves.

The purpose of the shield was to prevent the earth and water from falling or rushing into the working compartment as the tunnel walls were constructed. It was essentially a control device, but it did not work well. There were several floods, one of which resulted in the death of six men. The sewage-laden water and soil of the Thames River also gave off poisonous and combustible methane gas, which often filled the working chamber, regularly endangering the workers.

The improved shield method suggested by Jacobs and Davies involved forcing a metal cylinder through the earth by means of hydraulic jacks. The cylinder is bisected by a metal wall or bulkhead that is pierced by openings and divided into several working compartments. The advance rim of the shield forms a cutting edge which digs into the earth, and the section to the rear of the bulkhead forms the "tail," within which successive rings of metal are erected as the shield moves forward. At the rear of the bulkhead, a hydraulically operated erector arm is mounted vertically on a shaft in the center of the bulkhead. This arm is used to place the metal plates that are bolted together to form the rings that constitute the tunnel tube. As rings are added, the tube is progressively lengthened. The jacks push against the last ring formed, forcing the shield forward.

At the rear of the compartment where the rings are assembled is another concrete bulkhead pierced by several air locks. The earth and rocks not pushed aside as the shield advances are allowed to pass through openings in the first bulkhead and then are shoveled into small carts on rails that are then passed through some of the air locks. Other air locks are used by laborers, called "sandhogs." The air locks are necessary because the working compartment must be kept pressurized by forcing enough air into it to keep the mud and water outside from rushing into the compartment. It was the lack of a compressed-air working chamber in the Thames Tunnel which led to repeated flooding on that project.

When Haskins began the first Hudson River tunnel, he made an error in judgment that was essentially the reverse of that made by Brunel. Haskins believed that the material through which the tunnel would be dug was firm enough that he could forgo the use of a shield, instead filling the tunnel with compressed air to support the tunnel walls and to keep water out. He would then erect a thin shell of iron and line that shell with brick. He was mistaken about the strength of the river silt, and the tunnel flooded repeatedly. It was not until a shield was employed in 1890 that the work could continue safely.

Jacobs and Davies had used the shield method, with compressed air, in all the tunnels they had previously designed. In their opinion, it was the best method to use under the Hudson River, even though the cost of employing sandhogs was high because of the dangerous and physically demanding nature of their work.

There was a potentially cheaper method of tunnel construction, which some engineers later advocated for the Hudson River tunnel: the trench method. The method involved dredging a trench in the bed of the river and then towing prefabricated tunnel sections out into the river and lowering them into the trench from a floating facility. This method worked best when the bed of the river was fairly stable, thus lessening the possibility that the side slopes of the trench would collapse before the tunnel sections could be placed into position. It was also important that river currents were weak in order for the floating facility to be maintained in a stable position. Most important, this method was best employed in rivers with limited boat traffic in order to prevent collisions.

The trench method was used, to a limited extent, for the 149th Street subway tunnel under the Harlem River between Manhattan and the Bronx (it now carries the 2 train). Completed in 1905, this tunnel was built by a combination of methods. A dredged section with slopes stabilized by sheeting was covered by a timber roof, and the lower portion of the tunnel under this roof was built using compressed air. The Lexington Avenue subway tunnel under the Harlem River was also built by the trench method and completed in 1918 (it now carries the 4, 5, and 6 trains). For this tunnel, steel forms were lowered into a trench and then covered with concrete. Water was then pumped out of the tunnel and the interior lined with concrete under normal air pressure.[9] Both of these short tunnels were constructed in shallow trenches and in a slow-flowing river with limited boat traffic. The material excavated was a relatively stable mixture of silt, sand, gravel, and clay. Completed in 1910, the 8,390-foot-long Detroit River Tunnel linking the Michigan Central Railroad lines between Detroit, Michigan, and Windsor, Ontario, was the longest tunnel ever built by the trench method, and it is generally considered to be the first true subaqueous trench tunnel.

Jacobs and Davies did not believe the trench method was suitable for use in the Hudson River because of an unstable mix of mud and silt in the riverbed and because of the hard-to-excavate reef of solid rock near the Manhattan shore. The river was also heavily used by watercraft, which were likely to pose a threat to, and to be threatened by, a floating facility operated in the navigation channel of the river.

In regard to the need for ventilation, a concern expressed by many observers who thought that noxious gases given off by internal combustion engines would poison those who used the tunnel, Jacobs and Davies did not see this as an insurmountable challenge. They pointed to London's Blackwall (1897) and Rotherhithe (1908) tunnels, which, though built for horse traffic, were used by motor vehicles. These tunnels, however, were not good examples to demonstrate the viability of vehicular tunnels. They were much shorter and carried far less traffic than what was projected for the Hudson River tunnel, and therefore they could not suffer the same potential buildup of polluted air. But as their traffic levels increased, it was found that without a mechanical system for ventilation, they were very restricted in the number of vehicles they could accommodate. Eventually, they had to be retrofitted with mechanical ventilation systems. Perhaps in realization of this limitation, Jacobs and Davies proposed ventilating plants at each end of the Hudson River tunnel to supply fresh air and to exhaust foul air through ducts running the length of the tunnel.[10]

The total cost of the tunnel designed by Jacobs and Davies, including labor and materials, easements or rights of way, carrying charges, engineering, and contingencies, would be approximately $11 million. This, or course, was considerably less than the $42 million cost estimated by Boller, Hodge and Baird for a suspension bridge.[11]

Following release of the engineering reports, the bridge and tunnel commissions could no longer support plans for immediate construction of a bridge, given the comparative cost estimates and the urgent need to build the river's first vehicular link at a point where it was most needed. It seemed, therefore, with the issue of what type of structure to build resolved, that there might soon be real progress toward actual construction. But there was a hitch.

The constitution of the state of New Jersey would not permit appropriations of more than $100,000 without a referendum, and other laws stood in the way of the project. So the New Jersey legislature passed an act on April 17, 1914, allowing any three or more counties whose territory was contiguous and one of which was bounded by a navigable river adjoining another state to form a bridge and tunnel commission. Bergen County passed the appropriate resolution the following August, and Hudson County agreed in April 1915, but no other county would go along. Without appropriate action by the required three counties, the New Jersey legislature refused to authorize or fund a New Jersey commission from 1914 to 1916.[12]

In 1914, the New York State Bridge and Tunnel Commission recommended the immediate construction of a tunnel as soon as a means of providing the necessary funds was identified. But the New York commission could not act on its own, so little was accomplished during the following few years.

There was another organization at work, however, with its own plans and its own timetable. The Public Service Corporation of New Jersey (PSC) was an amalgamation of several hundred New Jersey gas, electric, and transportation companies, including most of the state's streetcar lines. In a 1912 report, PSC Chairman Thomas N. McCarter, a former New Jersey state senator from Essex County and former attorney general of New Jersey, stated that the corporation served 193 municipalities with one or more classes of utility and served most of them with all three: gas, electricity, and transportation. The Riverside and Fort Lee Ferry Company, which operated the heavily used Edgewater Ferry between New Jersey and 130th Street in uptown Manhattan, was just one of the transportation companies owned by the PSC.[13]

McCarter first called the attention of his executive committee to the benefits that would accrue to the PSC's service area by construction of a Hudson River vehicular tunnel in November 1913. Then, for the next three years, he watched and waited for something to be done. A graduate of Princeton University and the Columbia University Law School, McCarter left behind a promising career in law and politics to form the PSC in 1903 when he was thirty-six. As the *New York Times* observed years later on the event of his death, the history of the PSC and McCarter's career were inseparable. Throughout his forty-two years as president and chairman of the organization, he was a tireless advocate for private ownership of utilities, believing publicly run agencies to be too inefficient and, often, too corrupt.[14] Although he was known to be skilled at political negotiation, McCarter was averse to any compromise of his principles for the sake of political gain. As he later stated, on the day of his retirement, "From the beginning, I charged myself with the responsibility of being a trustee to see that the public got fair rates; that our employees were fairly treated; and that our security holders, past and present, should receive a fair return. We have never lowered that banner."[15]

On January 18, 1916, McCarter reported to the PSC executive committee that although the Hudson vehicular tunnel project had been discussed as a public enterprise for ten years, there was no immediate prospect for its construction. Therefore, he proposed, the PSC should study the subject to see if it was practical for the corporation to build it. The executive committee subsequently appointed a special committee to report on the feasibility and probable cost of such a tunnel. Needing the advice of technical experts, the special committee appointed a three-member board of consulting engineers. William H. Burr, Ralph Modjeski, and Daniel E. Moran composed the consulting board. Burr was a professor of civil engineering at Columbia University, a consultant to various municipal departments of New York City, and a specialist in bridge engineering. His name was also well-known to George Goethals, due to his service on the first (1899–1901) and second (1904–1916) Isthmian Canal Commissions, created to oversee construction of the Panama Canal. Modjeski, a consulting engineer with offices in Chicago and New York, was also a specialist in bridge engineering and one of the best in the country. Within a few years, he was selected to serve as chief engineer for construction of the Delaware River Bridge. Moran owned a New York City consulting engineering firm that specialized in structural foundation design. He also knew quite a lot about the working conditions that could be encountered in subaqueous tunnel construction, having patented a caisson air-lock design in 1893.

Over a period of about a year, these men made an exhaustive study of all relevant issues, including the problem of ventilation. Unlike Jacobs and Davies, they did not dismiss the matter as easily resolved and noted that "the precise degree of vitiation of the air of a vehicular tunnel before it becomes unsafe to breathe has not been determined." They also stated that air in the Blackwall and Rotherhithe Tunnels "has now become so vitiated that it appears probable that resort must soon be made to artificial ventilation."[16] In order to solve the problem of how much mechanical ventilation would be required in the Hudson River tunnel, they built a 125-foot-long, air-tight test chamber on PSC property in Newark, New Jersey, with a transverse cross-section almost identical to that which they proposed for the tunnel. The chamber included a roadway, sidewalks, air ducts at top and bottom, and electrically operated blowers and exhaust fans. Inside they placed eight automobiles and conducted tests with the motors racing and at idle, with the ventilation on and off, and with the direction of air flow reversed. Air samples were taken under various conditions of operation and the results analyzed. Their conclusion: "It is entirely feasible to ventilate satisfactorily such a tunnel tube when used by motor vehicles in numbers practically equal to its capacity."[17]

With a similar level of effort as that expended by the consulting engineers, PSC Secretary Percy Ingalls made a study of vehicular traffic carried by the existing ferry lines, with an emphasis on the ferries serving Lower Manhattan. He concluded that there was so much pent-up demand, particularly from trucks (both horse-drawn wagons and motor trucks), that even at a toll rate greater than that charged by the ferries, the tunnel would easily turn a profit.

In consideration of the report produced by the board of consulting engineers and the investigations conducted by Ingalls, the special committee made a recommendation to the executive committee for construction of a single-tube tunnel, with an internal width of twenty-five feet and two traffic lanes, one for eastbound vehicles and the other for westbound vehicles. The section of the tunnel that would go under the river would be constructed on land, then towed out and sunk in the riverbed in a previously excavated trench. The location would be the same as that recommended by Jacobs and Davies: between Canal Street in Manhattan and Twelfth Street in Jersey City. The total cost of the "fully equipped" tunnel, at the current prevailing wartime rates for labor and materials, including contractor's profit, engineering, and contingencies but not including easements and real estate for entrances, would be approximately $8.9 million. At "normal" labor rates and material costs, the cost would be approximately $6.9 million.

The board of consulting engineers made its report to the executive committee (but not to the public) on January 30, 1917, not long after Governor Edge returned to Trenton from his dinner with Goethals in Washington, D.C. Now, it was Edge's turn to receive a request for a conference. PSC vice president Edward W. Wakelee telephoned Edge and asked for a meeting with the governor. Wakelee and McCarter knew that the Board of Chosen Freeholders of Essex County had recently committed to joining with Hudson and Bergen counties to petition the legislature for authority to build a tunnel. But the necessary resolution and appropriation of funds could not be made until May 1917, when the county was scheduled to adopt its annual budget.

Because the state senate had to approve both the pending three-county petition and Edge's appointment of a new state tunnel commission, Wakelee and McCarter feared that the legislature would not be able to act until 1918, unless Edge was willing to call a special session. They did not think that the long-delayed tunnel project could wait that long. The PSC's executive committee was about to officially release the results of its year-long study of the proposed tunnel project, and the members of the committee were leaning toward a recommendation that the corporation be allowed to build it.

Early one morning, probably in February 1917, Wakelee and McCarter walked into Edge's office, shook hands, and sat down. McCarter then got right to the point. He wanted the PSC to build the tunnel. Although Edge gave the offer due consideration, and the men discussed the details of such a plan at length, he knew that to accept would endanger the Delaware River bridge project. Both initiatives were integrally tied together by the funding mechanism he had proposed. In rejecting McCarter's offer, however, he emphasized a different concern, telling his guests that if the state built the tunnel, the margin of profit that would otherwise go to the PSC could be eliminated to the advantage of the taxpayers. "The best interests of the state demand that we go through with the legislation that has been suggested," he concluded. With that, Wakelee and McCarter departed "with every evidence of good feeling," according to Edge.[18]

In the last week of March, the PSC released its official and voluminous 138-page tunnel report. The document stated that the tunnel could be built by the PSC, which would issue bonds to fund the up-front costs, with reimbursement via tolls or by the government as a toll-free facility. The important thing was that it be built as soon as possible. "The question [of] by whom it should be constructed," the report stated, "is of lesser importance." But the report also acknowledged that "the benefit derived . . . would be greater if it were operated as a free highway than if tolls were charged."[19]

Edge, of course, also wanted to build the tunnel as soon as possible. But before he could revive the moribund New Jersey commission, he had to make certain that the New York commission would cooperate. And that cooperation was in doubt, because of a monumental fight then being waged by the two states over shipping rates charged by the railroads. In May 1916, an entity with the unwieldy name of Committee on Ways and Means to Prosecute the Case of Alleged Railroad Rate and Service Discrimination at the Port of New York submitted a brief to the Interstate Commerce Commission (ICC) urging the ICC to require a change in shipping rates charged by the sixteen railroads then operating in the Port of New York. The problem, from New Jersey's perspective, was that the railroads provided free lighterage service between New Jersey and New York but charged shippers the same rates for shipments to or from New Jersey as paid for shipment to or from New York, even though there was no lighterage expense involved when the final destination of the freight was New Jersey. This was the basis of the infamous "New York Harbor Case," which Goethals biographer Joseph Bishop called "the culmination of two and a half centuries of bickering across the imaginary line dividing the harbor against itself." Bishop correctly characterized the case, "which began in bitterness, with sharp criticisms and savage sneers hurled back and forth by newspapers and civic organizations on both banks," as nothing less than a "political and economic war."[20]

The ICC had yet to issue its ruling when Edge began his campaign to soothe ruffled feathers and convince old enemies that their common interest lay in compromise and accommodation to the needs of others. On March 1, 1917, he delivered a speech at the annual meeting of the Chamber of Commerce for the State of New York, with representatives of business groups and public officials from both states on hand. "We cannot grow tightly bandaged in the swaddling clothes of provincialism," he told his audience. "We cannot progress while tied to the stake of territorial selfishness. Let us cut the rope and, as free, broadminded factors, do something for the mutual benefit of New York and New Jersey and the nation." Then he made a pitch for the tunnel project, saying, "When the project of a traffic tunnel under the Hudson was first proposed, the enthusiasm of its advocates was dampened somewhat by assertions that New York City would never cooperate in an improvement calculated to divert its population and its business interests." He was happy to say, he claimed, that New York had not assumed such a narrow-minded stand and that with the continued cooperation of New York and the northern counties of New Jersey the tunnel "ought to be an accomplished fact within a comparatively few years."[21]

Mindful of the fears of people in both states regarding the economic impacts of a transriver connection, Edge claimed that "population and business will not be diverted from New York to New Jersey nor from New Jersey to New York." As proof of this, he pointed out that construction of bridges across the East River between Manhattan and Brooklyn had benefited both boroughs, which had been separate cities when the first East River connection, the Brooklyn Bridge, was completed in 1883. "And so it will be with the traffic tunnel under the Hudson. Neither side can suffer. Both must feel the benefit of a progressive step which betters communication and improves transportation facilities," he urged. In foreshadowing the action that he intended his state to take, and with awareness that none of his plans would bear fruit without action on the other side of the river, Edge also stated, "It is unquestionably high time for New Jersey to awaken to its responsibilities, and I am making these statements in order that you in New York may see that at last New Jersey is beginning to perceive its responsibilities and its opportunities."[22] In conclusion, Edge called for creation of a joint bistate commission for the development of port and terminal facilities in New York Harbor.

Within the same week, Edge repeated his speech at a meeting of the New Jersey Chamber of Commerce, and he had the entire speech printed in the New York Times on April 15, 1917, so that the citizens of the city, including the boroughs of Brooklyn and Queens, could evaluate his argument for themselves. He knew that the greatest opposition to the tunnel came from those in New York who had long assumed that the current railroad-dominated system for the movement of people and freight was to their advantage.[23]

A meeting called by the Federated Civic Associations of New York, held March 12, 1914, served to illustrate the deep division of opinion that still existed about the issue of cooperation between the two states. Representatives of approximately thirty civic organizations argued for nearly three hours whether the federated organization should officially support improved transportation facilities, either bridges or tunnels, between Manhattan and New Jersey. Those speaking for Manhattan merchant associations were in support, while those representing real estate and taxpayer associations were opposed.

A delegate from the United Real Estate Owners' Association, Dr. Henry W. Berg, opened the debate by stating that the construction of the Pennsylvania Railroad tunnels in 1910 had been a mistake because the city could not toll the freight and passengers brought in by that means, and that New York would be robbing itself of its commercial supremacy by granting another state a further opening. In Berg's opinion, it was no wonder that New Jersey might favor improved facilities because it would receive all the benefit in the form of overflow population. Moreover, improved facilities would increase the value of New Jersey real estate five times and bankrupt Long Island and Bronx property owners.[24] Berg's comments were strongly endorsed by other real estate professionals present and were just as strongly refuted by others. Considerable rancor was expressed on both sides of the issue until the meeting adjourned. With conflict such as this among powerful business interests, it was no wonder that by 1917 little had been accomplished to advance a specific plan.

Nonetheless, Edge managed to win passage in New Jersey of the necessary legislation to fund the highway program, although a similar tax for the Delaware River bridge and Hudson River tunnel projects did not become operative until 1922, at the conclusion of the five-year taxing period for the highway program. It was Edge's plan that as soon as the Pennsylvania and New York legislatures took action to fund their part of the interstate projects, he would call a special session of the legislature to bring the bridge-and-tunnel tax forward for immediate levy.[25]

Edge also managed to secure passage of legislation to create the position of state engineer and to do so in such a way that Goethals, as suited his temperament, could run the highway program in the way that he saw fit. On Thursday, March 29, 1917, Goethals officially accepted his appointment as state engineer, and on the following Wednesday he began an inspection tour of the state highway system with members of the State Highway Commission. Two days later, the United States formally declared war against Germany. Eleven days after the declaration, while in Trenton for a meeting of the highway commission, Goethals received a letter from President Woodrow Wilson asking him to accept a position as general manager of the United States Emergency Fleet Corporation. His task would be to oversee construction of a merchant marine fleet of one thousand wood ships to carry troops and supplies to Europe.

As a soldier, albeit retired, Goethals could not turn down the call to duty in time of war. He asked Edge to relieve him of his responsibilities as state engineer. Instead, Edge worked out a compromise whereby Goethals could oversee the highway program, with the help of able assistants of his choosing, while also managing the shipbuilding program. The key to the compromise was Goethals's agreement to accept salary only for those days that he actually spent working for the state.[26]

On June 8, 1917, after having been duly petitioned by Bergen, Hudson, and Essex counties for the creation of a Bridge and Tunnel District, Edge appointed ten members of the newly named Hudson River Bridge and Tunnel Commission and directed it to convene a series of joint sessions with the five-member New York Bridge and Tunnel Commission to agree on specific plans for construction and financing. Thomas McCarter was appointed Commissioner at Large.

The three petitioning counties also appropriated approximately $30,000 for the expenses of the commission, which promptly retained Goethals to make a study of the Jacobs and Davies plan and the PSC plan and recommend adoption of a specific plan, which might not be either of the two already proposed.

The New York legislature appropriated another $5,000 for the expenses of its commission, which had been somewhat active conducting river-bed borings and making cost estimates, in cooperation with "interested New Jersey individuals."[27] From this point forward, the two commissions began to act less independently and more as a joint commission, with the chairman of the New York commission serving as acting chairman of the united bodies.

About this time, the commissions were also charged with considering interstate construction of tunnels for rail as well as vehicular traffic. At a recent ICC hearing on the New York Harbor case, great emphasis was placed on the potential benefits that might be realized in Lower Manhattan from constructing railroad freight tunnels leading from the New Jersey railroad terminals. The congestion and inconvenience caused by lighterage, experts testified, would be greatly relieved by direct, underwater rail connections.

If much of the freight lightered by the railroads could be carried across the river by truck, however, there would be no need for a railroad freight tunnel. As the *New York Times* noted on June 9, 1917, "The plan of the New York Commission for a vehicular and pedestrian set of tunnels would mean that instead of loading at the crowded west side Manhattan docks, trucks could run over to Hoboken and load and unload at the terminals there, not so crowded. This would give great relief to the freight congestion situation on the lower west side."[28]

While the New York and New Jersey commissioners were reorganizing and getting back to work, on July 19, Goethals received a letter from President Wilson, restricting his powers as head of the Emergency Fleet Corporation of the United States Shipping Board. The general had clashed with Shipping Board Chairman William Denman, who was committed to building only wood ships. Goethals wanted to build steel ships, which the steel industry supported, and had met with steel-industry leaders without Denman's approval. Goethals believed that the supreme powers necessary for him to do his work, which he was sure were his to exercise, had been denied. "Centralized power is essential for rapid and efficient work," he wrote Wilson, in a letter of resignation the next day. This reaction showed certain aspects of Goethals's character that were to become manifest again in regard to the Hudson River tunnel project.

For the moment, however, Goethals did not want to look at plans for a tunnel. He wanted to regain his honor by going to war. On July 31, he met with Secretary of War Newton Diehl Baker and asked for duty, rebuilding the railroads of France, which were to be used by U.S. troops. He was turned down. Goethals thereupon returned to New York and rekindled his partnership in the consulting engineering firm, which soon changed its name to George W. Goethals & Company, with new offices at 40 Wall Street. It did not take him long to resume his former duties.

On August 1, the New York, New Jersey Port and Harbor Development Commission organized, and after meeting with Goethals on August 22, the commission appointed him as "consulting engineer." He agreed to serve until a report on harbor improvement could be produced. Throughout September, October, and November, Goethals kept busy preparing reports for the Port and Harbor Development Commission, for the New York and New Jersey bridge and tunnel commissions, and for the New Jersey State Highway Commission.

On December 14, Goethals submitted his official report, dated November 24, 1917, to the New York and New Jersey tunnel commissions at their office at 115 Broadway. He amended it on December 18 and again on January 21, 1918, but the essential elements of the report remained the same. Goethals rejected the two plans previously submitted by Jacob and Davies and by the PSC. Instead, he recommended a single-tube tunnel, forty-two feet in diameter, made of interlocking concrete blocks. His tunnel would have two levels, with three lanes of traffic on each level, the slower trucks using the lower level and the faster, lighter "motor cars" using the upper.

Acknowledging that the "insidious and deadly effects" of poisonous gases emitted by automobiles and trucks "are not to be depreciated," he cited the tests conducted by the PSC and the "opinions of engineers versed in the subject" as confirming his belief that a proper system of ventilation could be designed. "As a matter of fact," he stated, "the air can be withdrawn from and introduced at as many points along the line of the tunnel as desired, thus subdividing the structure into a number of small sections."

The complete construction cost of the tunnel (as provided in his amended report of January 21, 1918) would be approximately $12 million. The real estate and easements would cost an additional $600,000, making the total estimated project cost approximately $12.6 million.[29] In his report, Goethals used a cost estimate provided by New York City contractor John F. O'Rourke. What Goethals did not reveal, either in his report to the commissions or in public, was that O'Rourke was more than just a contractor responding to a request for an estimate. O'Rourke was author of the design, and he probably approached the elder engineer soon after Goethals was appointed consultant to the state commissions. The entire course of O'Rourke's practice in New York indicates that had set his sights on being the general contractor for construction of a Hudson River vehicular tunnel whenever such an opportunity arose.[30]

Born in Tipperary, Ireland, on October 3, 1854, O'Rourke came to the United States with his parents when he was two years old. He attended elementary school in New York City and studied engineering at the Cooper Union for the Advancement of Science and Art, graduating in 1876. Throughout his life he claimed to have been a professor of civil engineering at Cooper Union, but that institution has no record of him ever having been on the faculty.[31]

From 1887 to 1890, O'Rourke was the chief construction engineer of the Poughkeepsie Bridge over the Hudson River, located at a point roughly midway between New York City and Albany. At the time of its completion, this railroad bridge, about one and a third miles long including approaches, was considered one of the most significant cantilevered truss bridges in the United States, in terms of its contribution to engineering practice.

In the mid-1890s, O'Rourke became a partner in the Stephens and O'Rourke Engineering and Construction Company, which specialized in building demolition and foundation construction. Before the decade was over, he separated from Stephens to establish the O'Rourke Engineering Construction Company, and in 1899 he won the contract for foundation work on the City Island Bridge across the Harlem River, linking City Island and Pelham Bay Park at Rodman Neck, Bronx.[32] In 1901, he filed a patent, the first of many throughout his life, for an improvement to the methodology for subterranean or subaqueous foundation construction. In the same year, he also began foundation excavation for the New York Stock Exchange. That contract was followed by similar work for the Hanover Bank (1901), the Atlantic Mutual Addition (1904), the Altman Building (1905), and the Municipal Lodging House (1907), all in New York City.

In addition to foundation work for bridges and buildings, O'Rourke won contracts for numerous railroad or subway tunnel projects. In 1903, his company began excavation of the tracks and yards of the New York Central Railroad from Fifty-Seventh Street south to Forty-Second Street. This was an approximately $5 million contract, employing thousands of people. As this project was under way, O'Rourke also joined with Ernest P. Goodrich in construction of the Bush Railroad Terminal in Brooklyn. In 1904, O'Rourke won another big railroad job, the contract for excavation of the Pennsylvania Railroad tunnels under the Hudson River, and won additional work excavating the site for the Pennsylvania Terminal building in New York City (Penn Station).

On February 6, 1912, O'Rourke filed a patent for a method of tunnel construction that included a concrete-block outer ring, and the patent (no. 1,043,348) was granted November 5, 1912. This was followed by other filings refining the design, including one on November 17, 1915 (no. 1,235,233), granted July 31, 1917.

From 1916 through 1917, O'Rourke served as vice-president and chief engineer of the Flinn-O'Rourke Company, which built the Old Slip–Clark Street Tunnel and the Whitehall–Montague Street Tunnel under the East River. There was another firm in existence at this time, the Booth and Flinn Company. Apparently, therefore, the Flinn-O'Rourke Company was a joint venture created solely for the purposes of constructing the two subway tunnels.

Months after Goethals released his report (including revisions), the *Engineering News-Record* published an article by O'Rourke (on March 12, 1918) in which he provided details and illustrations of his design for a shield-driven Hudson River tunnel, as adopted by the bridge and tunnel commissions and recommended by Goethals. He also mentioned that "application has been made for a patent on this shield construction."[33] He had actually filed two related patent applications in the same month. On February 1, 1918, he filed an application for a multiple pressure shield (no. 1,277,107, granted August 27, 1918), and on February 4, 1918, he filed an additional application for a tunnel shield (no. 1,296,312, granted March 4, 1919).

Although the tunnel design submitted by Goethals later came to be informally referred to as the "Goethals-O'Rourke" plan, when Goethals released his report at the end of 1917, many people assumed that he was the primary designer. O'Rourke initially did not seem to care that Goethals adopted his design and presented it as his own. His sights were set on the money that would come his way via the construction contract or by sale of his patent rights. And at the time, the source of the design probably did not matter much to the bridge and tunnel commissioners. What mattered was that they had a plan that they could present in order to secure enabling legislation.

On December 15, 1917, Goethals also submitted his annual report to the New Jersey State Highway Commission, stating that improved roads were never before so greatly needed in New Jersey, as the state was "one of the main entrances to the country's outlet, the port of New York." With a veiled reference to the need for a vehicular tunnel across the Hudson River, he also said, "The congested condition of the railroads for some time past brought the motor truck to the fore as a means for transporting farm and manufactured goods." And he made a strong recommendation for concrete as the best material for withstanding the heavy loads to be borne by trucks using the new routes to be established.[34]

While Goethals was in the process of amending his report to the state bridge and tunnel commissions, on December 18, 1917, he was recalled to active military duty as Acting Quartermaster General of the Army. Despite the new demands of that position, he continued his association with the engineering consulting firm that bore his name and began his service as president of the Wright-Martin Aircraft Corporation, a position he was elected to the previous October. One of his engineering partners, George H. Houston, was vice president and general manager of the company, and it was he who probably arranged Goethals's involvement. The company manufactured the Simplex automobile and aircraft engines for the French government. Earlier in the year, it had received a large order from the U.S. government for manufacture of approximately four thousand aircraft engines. These motors were to be manufactured at the company's New Brunswick, New Jersey, plant.

With submission of Goethals's report to the state bridge and tunnel commissions, it appeared that real progress was being made toward adopting a set of plans so that construction contracts could be issued. This appearance was deceiving. On December 17, 1917, the New York Bridge and Tunnel Commission appointed New York attorney Paul Windels as counsel to the commission. Windels soon informed the commission that New York State could not pass any appropriation exceeding $1 million without a referendum. In addition, for the two state commissions to be able to issue construction contracts, they would first have to contract with each other to manage the project. Such a contract constitutes a treaty, and the Constitution of the United States does not allow states to enter into treaties without the consent of Congress.

The complications came in triplicate. First, there was the need for a referendum; second, a treaty between the states needed to be negotiated and approved by Congress; and third, on the same day Windels became counsel to the New York commission, the ICC decided the New York Harbor case (although it was not officially handed down until January 22, 1918), saying that the New Jersey railroads should be extended under the Hudson River to Manhattan.

The ICC decision compelled the two states to work together in a spirit of cooperation, but not for construction of vehicular tunnels. The ICC clearly recognized that the existing methodology for handling freight in the port area needed to be "thoroughly revised," but the revision suggested by the commission depended on extension of the rail system of New Jersey into New York via tunnels, and not on extension of the highway system. "Adequate freight tunnels under the North River," the ICC stated, "would make it possible to handle a large portion of Manhattan's freight traffic without the use of lighters or floats." The decision also found that with creation of a freight-handling system involving subaqueous rail tunnels and new terminals, "the congestion on the west side of Manhattan Island caused by the assemblage of countless vehicles at the countless piers to receive and discharge freight would be considerably relieved."[35] What the ICC decision apparently did not anticipate was that the thousands of rail cars brought into Manhattan each day would still have to be unloaded at terminals and that trucks would still congest the city streets as they congregated at these terminals to receive or deposit their loads.

The tunnel commissioners, of course, had little interest in building a tunnel for rail traffic. Furthermore, planning for a rail-freight tunnel necessarily included planning for new terminal facilities in Manhattan, and such plans would take years to work out. Unless the necessity of immediate action toward construction of a vehicular tunnel could be made apparent, the commissioners might find the political will for such action lacking. As 1917 drew to a close, the greater considerations of how to solve the problem of freight movement in the Port of New York threatened to indefinitely delay progress toward construction of a vehicular tunnel. To Walter Edge and many of the tunnel commissioners, it may have seemed that only an act of God could move the project forward.

3

A Coal Famine

Twelve months ago no one would seriously have urged that the city could be in the grip of a coal famine involving in its financial features alone the loss of many times the total cost of construction of the proposed tunnel, and that this could be largely due to its inability, because of the ice-choked river, to transport thousands of tons of coal that were literally within sight of its inhabitants and yet as unattainable as if they were still in the mines.

—New York State Bridge and Tunnel Commission, *Eighth Report of New York State Bridge and Tunnel Commission to the Legislature of the State of New York, 1918*

THE HIGH TEMPERATURE in New York City on Friday, December 28, 1917, reached just twenty-seven degrees Fahrenheit. It took just five hours for that number to drop to eight degrees. By 8 a.m. on December 29, the temperature had plunged below zero and was continuing to fall.[1]

A fast-moving cold front of unprecedented severity had blasted the entire northeastern United States, bringing unusually high atmospheric pressure and frigid temperatures to an area stretching from the Upper Mississippi Valley to New England. In the New York City metropolitan area, the size and weight of the mass of high-pressure cold air quickly pushed the lower-pressure, warmer air out to sea.

At that time, coal was the primary fuel used to produce heat for buildings, power for public utilities, and electricity for the subway system in New York City. As temperatures dropped, demand for coal increased. Homes, apartment buildings, settlement houses, retail businesses, offices, schools, hospitals, courts, prisons, factories, and other public and private institutions all needed more coal and soon.

Most of the city's coal supply arrived on barges, with lesser amounts brought in by car float, and some by truck. A relatively small amount was brought in directly by rail. The barges were loaded with coal at railroad terminals in New Jersey and then towed by tugs across the Hudson River or through the bay and up the East River to unloading terminals in Brooklyn. For several days before the cold front descended, there had been a substantial reduction in the amount of coal delivered to New Jersey's railroad terminals. The reduction was due partly to holiday-related work breaks in the coal mines and partly to three significant problems of the railroads: a shortage of coal-carrying cars, frozen switches, and snow-covered tracks. There was coal at the terminals—much of it sitting frozen in rail cars waiting to be unloaded or mounded in frozen piles on the ground waiting to be loaded onto barges. On a typical winter day, the city consumed between forty thousand and forty-five thousand tons of coal, representing from nine hundred to one thousand car loads. On Friday, December 28, 1917, only about two-thirds of the normal daily coal shipment was delivered.[2]

The city had already experienced the repercussions of coal shortages earlier that month on December 12, when riots erupted in several parts of the city. The coal the rioters desperately needed, forty thousand tons of it, had been delivered to New Jersey terminals that day, but little of it made it to New York City. Small disturbances occurred at the fuel sellers' storage and distribution yards in the Brownsville section of Brooklyn and at the Rubel Brothers coal yards in East New York. The most serious incident happened when an estimated crowd of two thousand people, mostly women, some with children in their arms, stormed the yards of S. Tuttle's Son and Company in Brooklyn's Williamsburg section. People hurled loose bricks and stones they picked up off the ground, and women threw the dishpans and scuttles they had brought with them to carry coal. They aimed the missiles at the shop's wagons, windows, and even employees. Although no one was seriously hurt, several rioters were knocked down and trampled by the mob, and many windows were broken before police arrived and dispersed the crowd.

In response, Tuttle announced that as soon as his company received a new consignment of coal, he would distribute it to people in small lots instead of selling it in bulk to industrial or institutional buyers. But there was a limit to what Tuttle and other fuel distributors could do. Not only were they subject to the weather, but due to the nation's involvement in the First World War, coal distribution was controlled by a hierarchy of federal, state, and local fuel administrators. Although officials in that three-level hierarchy knew that an impending fuel crisis was at hand, they, too, were limited in what they could do.

Even as Federal Fuel Administrator Reeve Schley announced that an extra twenty thousand tons of coal had been diverted to terminals in New Jersey, he made it plain that New York City was not out of danger from the shortage. As the *New York Times* reported on December 13, "fuel conditions were acknowledged in local official circles yesterday to be rather discouraging."[3]

Over the following weeks, weather conditions worsened, and it became increasingly apparent that the delivery of more coal at terminals in New Jersey would not alleviate the coal shortage in New York City. The Lehigh Valley Railroad unloaded 202 railcars, about ten thousand tons of coal, at its Perth Amboy terminal in the twenty-four-hour period ending Saturday morning, December 29. Approximately fifty thousand tons were waiting to be moved from that one terminal alone.[4] At other New Jersey terminals, thousands more tons of coal sat, ready to be transported into New York. But there were not enough men available to thaw the coal by spraying it with steam, to break it up, and to load it onto barges. This was due partly to manpower shortages caused by the war and partly to the unwillingness of potential workers to take on the grueling and low-paying work in such frigid conditions. Even in good weather, the job of unloading coal from rail cars and reloading it onto barges was hard physical labor, and the laborers available to do the work were often unreliable. Once they accumulated a little bit of cash, they often quit.[5]

The coal that could be loaded onto barges still had to be towed across the Hudson River to Manhattan or across the bay to Brooklyn or up the East River to the Bronx or to Queens. The Philadelphia and Reading Railroad and the Pennsylvania Railroad owned their own tugs, the biggest and best in the harbor, but the other New Jersey railroads relied on more than a hundred independent tug companies. The efficiency of those independent tugs in moving coal depended on a reliable labor supply and on free navigation of the river, unimpeded by ice and fog.

As temperatures fell during the last few days of 1917, the buildup of ice in the Hudson River gradually increased. Most years, the point at which the river froze solid from bank to bank was well above the northern tip of Manhattan Island; but free-floating ice could present problems even in a winter of moderate severity.

As the tide flowed in, the tugs could better resist the flow of ice as they steamed upriver; as the tide flowed out, however, navigation could be perilous even as far south as Lower Manhattan. As the temperature dropped and ice accumulated, some of the tugs began to suffer damage to their propellers from striking floating ice, which put them out of service.[6]

Many New Yorkers still remembered the winter of a few years before, when the North River froze down the entire west side of Manhattan, and no tugs or barges could move at all. At that time, no coal famine resulted because land value was low enough for the maintenance of large coal-storage yards in various parts of the island. Just after the turn of the century, there had been 412 yards in Manhattan; in 1917, there were only 63. Not only did this leave the island with nearly 80 percent fewer coal-storage facilities, but the average size of the remaining yards was generally smaller than in previous decades. Therefore, the island's entire storage capacity in 1917 equaled only about three or four days' supply, at most. Many hotels, offices, and institutions had long ago adopted the practice of buying coal on a day-to-day basis or every few days. On any given day, the amount of coal that could be stored in Manhattan was significantly less than that stored in rail cars, in piles on the ground, or in barges in New Jersey.[7]

Knowing that coal supplies were perilously low with severe cold weather on the way, New York State Fuel Administrator Albert H. Wiggin rushed back to New York from Washington, D.C., on Thursday evening, December 27. On Friday, he met all day with managers of light, heat, and power plants to work out a program of strict energy conservation. The meeting was reactive instead of proactive, however, and it took place too late to avert the crisis. Public officials knew the city was vulnerable, even before the bad weather hit, and their worst fears were soon realized as the punishing weather exacerbated the shortcomings of the city's system of coal distribution. The drama of collective suffering that was to play out over the following few days was revealed in numerous small incidences that occurred throughout the city.

On Saturday morning, December 29, Brooklyn doctor Ira Cohen received a summons from a distraught mother, whose two-month-old infant had suddenly taken ill. By the time Cohen arrived at the unheated flat of the Daldananus family at 72 Union Avenue, the child was already dead. When Coroner E. C. Wagner arrived, the child's father, Joseph, told him that he had spent most of his money for food. When the cold front suddenly blew in, the small amount that he had set aside for fuel was of no use because there was no coal to be bought. His baby, who had previously been in good health, quickly succumbed to the cold. Not wishing to see the other two Daldananus children suffer, and perhaps die, Cohen and Wagner arranged for some coal to be brought to the flat.

Approximately one-half mile northeast of the Daldananus home, hundreds of men and women, many with children in tow, withstood a thirty-mile-per-hour wind that drove light, dry snow into their faces as they con-gregated at the big coal yards on Newton Creek, between Grand Street and Montrose Avenue. They were there because, like Joseph Daldananus, they had found that the smaller retail coal dealers had nothing to sell, and they believed that some supply might be available in the yards of the larger dealers. But there was none. When informed that they could not buy coal at any price, many of the women in the crowd became desperate and agi-tated. Police had to be summoned from the nearby Stagg Street Station to regain order.[8]

Police throughout the boroughs of Long Island had their own problems keeping warm. North of Newton Creek, at the 247 Precinct station house at Grand Avenue and Crescent Street in Long Island City, the furnace went out for lack of fuel, and the officers on duty sat bundled up in their over-coats. Although they had ordered coal more than two weeks previously, none had arrived.[9]

In the Bronx, Deputy Fuel Administrator Joseph A. Hall had already placed forty-five apartment houses on a list for emergency delivery of coal.[10] As in the other boroughs, thousands of people went out, walking the streets searching for fuel and pleading with coal-yard operators. During the day, more than five hundred people went to the Bronx fuel administra-tor's office to demand that something be done.

In Manhattan, the fuel shortage was even more extreme. Shortly after first light, scores of would-be customers, including children clutching pails, scuttles, and bags, began walking from place to place in search of fuel. Her-man Harjes, a coal dealer at 353 East 117th Street, normally sold most of his supply to bakeries, restaurants, or small coal dealers who then sold directly to the public. His winter sales usually averaged approximately 150 tons a day. On Saturday, he had only about five tons available when he opened his doors to a shivering throng of desperate customers, many of whom had already been to several other dealers to no avail. Unable to meet the needs of the crowd, Harjes allowed many of the coldest of them to come into his store, where they clustered around his steam heaters.

David Tropp, a dealer at 302 East 120th Street, received about thirty tons of coal on Friday, December 28. Instead of delivering it to his regular pa-trons, he doled it out in small lots to those who lived in his neighborhood. When word spread that another twenty tons were delivered to Tropp on Saturday, he was besieged by hundreds of people who had been unable to find fuel anywhere else. After Tropp ran out of coal, many of those waiting in line refused to believe that he had no more to sell. The crowd rushed into the store, looking for the coal they believed had to be there and threatening to wreck the place. The police were called, and when they arrived, there were still about a hundred people waiting, demanding to be let into the storage yard so they could see for themselves that all the coal was gone.

Fred Klobke managed the Central Coal Company, located at 155th Street and the Harlem River, which supplied many of the apartment houses in Washington Heights. All day on Saturday, he received calls from people who were out of fuel, but he could do nothing to help them because for the first time in years he was completely out as well.

One real estate agent divided apartment houses into three groups: those that had several days' supply of coal, those that were getting just enough on a daily basis to keep from completely running out, and those that had already run out and could not get more. By his estimation, the number in each group was about equal. Yet even in those buildings with coal, the supply was so limited that many families were closing off rooms and huddling around stoves in kitchens or heaters in dining rooms to conserve the coal they had. A representative of one large real estate firm that operated many apartment houses stated to a newspaper reporter that most of its properties were getting by on a day-to-day supply. "It is pretty bad now, but this is nothing to what it will be unless coal begins to move more freely,"[11] he said.

In the city's many settlement houses, the situation was equally dire. The Spring Street Settlement, which operated two houses, ran out of coal and had to move the children in its nursery to a nearby building at 224 Spring Street, which was heated. The Union Settlement at 237 East 104th Street and the Harlem Settlement at 405 East 116th Street had some supply remaining, and it took in all children, and a few parents, who had been forced to evacuate their unheated homes. When the New York Home for the Homeless Poor at 445 East 128th Street ran out of coal, manager Clinton Eva and his thirty-five boys went out and bought oil heaters. So many others were out doing the same, however, that an acute oil shortage developed.

The effects of the coal shortage were felt not just inside buildings. The Brooklyn Rapid Transit Company gave the public the option of either riding in unheated cars during certain hours of the day or having fewer trains in service. On Saturday, December 29, the company announced that it would not turn heat on in its cars during the morning and evening rush hours.

Public institutions also suffered. St. Vincent's Hospital in Manhattan called the fuel administrator's office to report that its coal bins were nearly empty and that William Farrel & Son, the hospital's regular dealer, refused to supply any more. A call from the administrator's office to the dealer resulted in an emergency delivery.

Patrick Jones, superintendent of supplies for the city school system, had been working night and day to increase coal supplies since schools closed for the Christmas holiday, but with reopening scheduled for January 2, he doubted that there would be enough. Forty schools had less than a ton each, and more than one hundred had less than a five-day supply.

One of the largest coal dealers, looking out his window at the falling snow, told a reporter for the New York Times, "If this keeps up we are going to have a real famine. I have been sitting here three days waiting in vain for coal. Even if we do get coal at tidewater [meaning at the coal terminals in New Jersey] it is going to take hours to thaw it out for loading onto the barges. The situation is the worst that I have ever known."[12] By the end of the day on Saturday, only about eighteen to twenty thousand tons of coal, less than half a day's supply, had been brought into the city.[13]

The following day, Sunday, December 30, the temperature dropped to thirteen below zero, the lowest temperature ever recorded by the Weather Bureau in New York City.[14] Because most businesses and public institutions were closed, many people were able to stay bundled up at home, lying in bed, fully clothed. Some residents of private houses and apartments, unable to stand the numbing discomfort of an unheated residence, left their homes to stay in hotels.[15]

As most people headed back to work on Monday morning, December 31, less than one day's supply of coal remained in the city. The temperature had risen, but only to seven degrees below zero, making it the second-coldest day on record in Manhattan. The big thermometer in front of Perry's Drug Store in Park Row, which recorded conditions at street level and which for many people was the most reliable indicator of true temperatures, registered two degrees lower than the Weather Bureau's instrument atop the Whitehall Building. As the New York Times reported, with some degree of understatement, "lower Manhattan awoke . . . to the realization that the coal situation had become acute."[16]

In many banks and brokerage houses in the Wall Street district, employees had to work in their overcoats, and in many buildings, elevators stopped running because there was not enough fuel to produce energy for their operation. At the Chase National Bank, 61 Broadway, passersby could look in the windows and see bundled-up employees in hats and overcoats hunched over their desks, struggling to grasp pens and pencils with numbed fingers. The fact that the bank's president, Albert H. Wiggin, was also the New York State fuel administrator, served to remind citizens that few could escape the consequences of the fuel famine.

At the Criminal Courts building on Center Street, the elevators stopped working at about 2 p.m. Shortly thereafter, the lights went out, forcing the district attorney's office staff to finish their work by candlelight. When the time came to go home, they used candles to light their way as they crept slowly down the stairs and out of the darkened building. At the Tombs, connected to the courts building on the Franklin Street side, the promised deliveries of coal had not arrived, and there was only enough in the jail's bins to last through that day. Warden John A. Hanley made plans to wrap all 450 inmates in blankets and to order them to bed if more fuel could not be found.

Along the waterfront in Manhattan, yet another effect of the coal shortage was evident. Scores of steamships scheduled to sail on Sunday or Monday sat idle in their berths, their bunkers empty. The cargo-laden boats had no fuel to fire their boilers. Economic losses were expected to be in the hundreds of thousands of dollars.[17] Soon, however, they might find movement impossible not only from a lack of fuel but also due to an abundance of ice.

The weather worsened, growing ever colder. The point at which accumulated ice impeded movement on the river gradually crept south. On Monday, December 31, it was reported that the Riverside and Fort Lee Ferry Company ferryboat running between Edgewater in New Jersey and 130th Street in Manhattan had been swept more than a mile downstream by ice floe. And a small tugboat became frozen in place at a point just above Dyckman Street.[18]

As the tide flowed in on the river, the ice floes were held back, making it possible to tow the coal barges to Manhattan. But as the temperature dropped and the ice floes increased, barge movement on the river would gradually slow. When the ice froze solid from bank to bank, all boat and barge movement would halt completely. Conversely, however, pedestrian movement across the river would increase.

On January 2, 1918, Frederick Gabay, a twenty-two-year-old deaf-mute employed by a munitions plant, walked across the Hudson River from Hastings, New York, to New Jersey, at a point just north of the city of Yonkers. In the middle of the river, he had to jump some bad spots, but he made it. It was the first time in five years that anyone had been able to cross the river on solid ice at a point so far south.[19]

With temperatures continuing to drop, New Yorkers enduring the cold weather inside their homes and offices began to wonder at whom they should direct their anger. Given widespread rumors of favoritism in distribution and price gouging by retail coal dealers, many citizens believed that these people, who were the links in the chain of delivery with whom they dealt directly, were at fault. Some favoritism and price gouging had, in fact, taken place, but that behavior occurred only once there was a shortage; it did not explain why the coal famine occurred in the first place.

Coal dealers and government officials tended to blame the railroad corporations. In their opinion, the railroads were responsible for seeing that the necessary commodity made it from New Jersey into New York. The railroads, for their part, claimed that they were responsible only for delivering the coal to New Jersey; moving it from there was up to the independent towboat operators. J. W. Searles, deputy commissioner of the Tidewater Coal Exchange, supported their stance, saying that the railroads' responsibility for coal delivery extended up to the placing of the coal into barges at New Jersey ports, "with the exception of the Reading and the Pennsylvania."[20]

Amid all the finger-pointing, there existed a pervasive and generalized feeling of helplessness that seemed to transcend individual or corporate responsibility. An agent at the downtown office of Burns Brothers expressed this feeling well when he told a crowd assembled outside his store, pleading for fuel, "I can do absolutely nothing for you. The railroads were snowed up, and the government cannot do more than it is doing. I am sure I do not know what we can do."[21]

In truth, heavy snow on tracks and frozen switches were making it difficult for railroads to move freight of any kind, but that alone could not explain the coal famine. The railroads were, of course, subject to delays caused by weather, but they were also extremely inefficient and uncooperative in establishing unified freight-terminal facilities for the New York metropolitan area. Each corporation acted in what it believed to be its own best interest and in opposition to the interest of its competitors or the larger community. But the railroads were only one part of the problem; the entire system of freight distribution was at fault, and there was little that the railroads would or could do in the short term to solve the situation. This was a crisis many decades in the making. There was, however, something more that government could do, and do immediately.

On Monday, December 31, Fuel Administrator Wiggin sent a telegram to Director General of Railroads William G. McAdoo in Washington, D.C., urging that the Pennsylvania Railroad tunnels under the Hudson River and the East River be used to run coal trains underneath Manhattan and over to Long Island for distribution to yards in Queens and Brooklyn. McAdoo, appointed to his position only days before, was well-known in New York City. He had been the primary force behind completion of the Hudson and Manhattan Railroad subway tunnels between New Jersey and Manhattan in 1908. Now, as the most powerful figure in the United States Railroad Administration, he exercised broad powers in execution of his responsibility to ensure efficient operation of the railroad industry during wartime.

"This matter is of the greatest urgency," Wiggin wrote McAdoo, "and this method will enable coal to be delivered to Long Island and Brooklyn from three to ten days sooner than if by regular transportation by river float. If [the] river freezes, as indications point, it will be the only way of getting coal to these sections."[22] And as the city's leading coal sellers had warned, if the Hudson River froze from bank to bank completely, using the tunnels might be the only way to keep thousands of people from death.[23]

The municipal authorities did not object to using the tunnels in the emergency, even though the city franchise granting the railroad the right to construct and operate the tunnels did not allow their use for freight.[24] The city also responded to McAdoo's call for city employees, mainly those in the street-sweeping department, to be employed in the railroad terminals to help unload coal from rail cars and load it onto barges or into the cars that would pass through the tunnels. But there were no places adjacent to the tunnel tracks in Manhattan where the coal could be transferred to horse-drawn wagons or trucks. Therefore, any coal supply destined for the island would have to be unloaded at the Long Island Railroad yard at Sunnyside in Queens and then trucked back into the city via the Queensboro Bridge for eventual delivery to the yards of the retail coal dealers.

As plans advanced for use of the Pennsylvania Railroad tunnels, the plight of people in all boroughs of the city worsened. Charity organizations, police stations, and hospitals treated hundreds of people suffering from the cold. The city health department reported on December 31 that the number of deaths and cases of pneumonia had increased dramatically within the last two days of the crisis. The coroner's office officially listed twelve deaths attributable to the frigid conditions. City department heads who were engaged in relief work joined with officials of charity organizations to warn that, unless the onslaught of cold ended soon, the number of deaths would undoubtedly increase rapidly, even when the supply of coal increased.[25]

On New Year's Day, a growing sense of panic and desperation swept through crowds gathered outside the retail coal yards of Manhattan. Several hundred men, women, and children stopped coal-laden wagons leaving Burns Brothers at 110th Street and East River. The fuel was intended for city institutions, but the mob unhooked the chutes on the wagons and then scrambled for fuel as the cargo spilled onto the street. Once again, as in so many instances in the preceding few days, the police were called to regain order.[26] Similar scenes played out in other parts of the city as growing desperation and anger escalated into mob action. Clearly, something had to be done, and done fast. Someone in a position of real authority needed to take forceful action. That someone was McAdoo.

On January 1, McAdoo took over active control of the Pennsylvania and the Long Island Railroads, ordering that shipments of coal be given preference over passenger traffic in the tunnels. Just after midnight the following day, the first coal-laden trains began to move through, each train consisting of about ten cars, and each car carrying approximately fifty tons of coal.[27] On the morning of January 3, the Pennsylvania Railroad hauled about seventy cars through from New Jersey to Long Island. This number was far less, however, than the one hundred cars it planned to send. Some of the fully laden cars were too large to pass through the narrow tunnels, which were not designed to accommodate freight.

None of the first coal shipments, though, made it to Manhattan. About half of the thirty-five hundred tons brought through was delivered to and used in Brooklyn, and the rest was distributed and used in Queens or in the numerous small towns of Long Island.[28] Even if the entire load had reached Manhattan, however, it would still not have been enough to meet the need. Even at full utilization, it might not be possible to send more than six thousand tons a day through the tunnels.[29]

As conditions worsened, other aspects of the city's inadequate freight-distribution system became apparent. Like the coal supply, most of the city's food had to be brought across the Hudson River, although food arrived not by barge but by ferry. With movement of all ferries held up by ice in the river, supplies of the most perishable food products, such as fresh milk (brought in by the railroads to New Jersey from as far away as Ohio and Michigan), began to run out.

On January 3, newly elected New York City Mayor John F. Hylan held a public hearing at City Hall, at which the complaints of approximately two hundred people, mainly women, were heard. One African American woman, Helen A. Holman, told Hylan that there were fifty thousand of her people hungry and cold in that part of the city north of Central Park. "They do not get even the coal doled out to people on the East side," she said. And in a patriotism-slanted appeal that might be expected to elicit some sympathy with a world war raging, she added, "The families of colored soldiers are suffering terribly."[30]

Mrs. Theresa Malkiel, speaking for a delegation from the Working Woman's Food and Fuel Committee, demanded that the city immediately open stations to sell coal and milk to the poor at cost. She claimed that poor people were freezing to death and babies starving for lack of milk. When another woman warned that there would be riots if something was not done, Hylan replied that he did not want to hear any more about riots. He did, however, issue orders to the police department to ensure that the distribution of coal was conducted fairly and that those people in line at coal yards would be served in the order that they arrived. How he intended to increase the supply of coal so that those standing in line could actually obtain some, he did not say. After about an hour, Hylan promised, "I will do all in my power to relive the suffering which you have described," and then terminated the meeting. Later in the afternoon, he appointed a Committee of Fuel Conservation, but he took no further steps to address citizens' complaints.[31]

The lack of immediate action by Mayor Hylan left few options available except curtailment of services. All of the General Sessions Courts in the Criminal Courts Building, except one, closed. Schools throughout the city also closed as their fuel supplies ran out. On January 4, the Metropolitan Museum of Art announced that it would close at night to conserve fuel. A week later, there were calls to close schools, banks, theaters, moving-picture houses, and restaurants three days a week. As might be expected, organizations representing restaurateurs, bankers, theater owners, and other types of business howled in protest. A group of theater managers and vaudeville operators asked leading theatrical producer George M. Cohan to send a letter to President Woodrow Wilson imploring him to prevent such action. Cohan wrote Wilson that the request for presidential intervention was not made with selfish motives but only because it was believed that the closure order would be selectively enforced, to the detriment of the theater owners and those whom they employed.[32]

An editorial in the *New York Times* may well have expressed the doubts of many people in the community: "Isn't the talk of rushing to Washington for orders to suspend industries, isn't the talk of 'closing New York for three days' a premature confession of despair?"[33] Yet, with people dying, offices and industries closing, and babies starving for lack of milk, those officials responsible for doing something no doubt felt that they had reason to panic.

President A. H. Smith of the New York Central Railroad, who served as assistant director general of railroads in charge of the eastern territory of the United States Railroad Administration, called on the United States Department of the Navy for an ice-breaking ship to open a channel in New York Harbor. He was told that there was no such vessel available. He then called the admiral in charge of the Brooklyn Navy Yard for help, but to no avail. When he asked relevant advisors if it might be possible to dynamite a path through the ice, he was told that such action was "inadvisable." With no other course of action available, he finally ordered W. B. Polluck, chairman of the Railroad Managers' Marine Department, to requisition an ice-breaking tug from each of five railroads: the New York Central, the Baltimore and Ohio, the Lehigh Valley, the New Haven, and the Lackawanna. These tugs were to come from points outside the Port of New York.

The tug from the Baltimore and Ohio did not show up on time, but the other four were joined together by hawsers on the morning of January 4 and used collectively to bash a hundred-foot channel through the Kill van Kull, the three-mile-long tidal straight separating New Jersey from Staten Island, thus freeing barges loaded with more than one hundred thousand tons of coal.[34] But those barges still needed to be towed through the harbor and up the Hudson and East Rivers.

As the struggle to bring fuel and food into the city continued, the inability of the state and local governments to quickly rectify the situation became increasingly clear. Despite the coal shipments through the Pennsylvania tunnels and the coordinated use of railroad-owned tugs to break up the ice in the port area, the daily amount of coal reaching the city was still far less than the forty thousand to forty-five thousand tons needed. On January 14, there were three hundred thousand tons of coal in New Jersey, but only about thirty thousand tons were delivered to New York City.[35] Like it or not, the only short-term solution to the problem of freight distribution would be an improvement in the weather that would both reduce demand for coal and begin to melt the ice in the waters of the port.

Warmer weather did return toward the end of January, and the crisis gradually abated as demand decreased and coal supplies returned to normal winter levels. But political leaders' intense frustration and citizens' widespread despair over their collective inability to transport essential commodities across the relatively short distance from New Jersey to New York created a deep desire to achieve a permanent, dependable solution

—a solution long contemplated but never realized. It was time to break through the political barrier still blocking construction of the Hudson River vehicular tunnel.

William M. Van Benschoten, chairman of the Commission on West Side Improvement, was one of the first New Yorkers to call for immediate action. On January 19, 1918, in a preliminary announcement of the commission's pending report to the New York State legislature (officially submitted on January 31, 1918), Van Benschoten warned that the current scarcity of coal, and the threat of a pending scarcity of food, brought into sharp focus the price that New Yorkers were paying for "having permitted politics, prejudice, personalities, indifference, and incapacity to prevent in the past the development of adequate and efficient freight terminals, as well as up-to-date and progressive transportation facilities between this city and the New Jersey shore."[36]

Citing the example of coal brought through the Pennsylvania Railroad tunnels, unimpeded by ice in the Hudson River, Van Benschoten asked, "What would have been the situation had there been tunnels connecting these thousands of tons of coal on the Jersey shore with Manhattan?" In answer, Van Benschoten stated that "the only efficient, at all times adequate and proper solution of the Jersey-Manhattan transportation problem is the tunnel, free as it would be from the fiercest storm or the coldest weather." To that end, the commission's report recommended that the legislature take immediate steps to have at least one tunnel built under the river for railroad or vehicular traffic. If the first tunnel was for the exclusive use of motor vehicles, the railroads could be expected to follow with construction of rail-only freight tunnels because they would obviously profit by being able to bring freight directly into Manhattan.[37]

The Commission of West Side Improvements, in its subsequent report to the New York legislature, also echoed the findings of the ICC in stating that new terminal facilities in Manhattan were necessary. Specifically, the report called for either a below-grade or an elevated terminal system on the west side extending south from Sixtieth Street to Canal Street or further, as conditions may require, with intermediate zone stations that all of the railroads would use. Despite the public comments of Chairman Van Benschoten on January 19, which seemed to indicate that it would be acceptable to the commission if a vehicular tunnel was constructed first, the report made it clear that a vehicular tunnel was not part of the proposed plan. The report also recommended creation of a Terminal Improvement Commission of seven members, consisting of the mayor and controller of New York City, two members of the Public Service Commission, and three members to be appointed by the governor. This commission would have the power to construct one or more rail tunnels under the river and to build a terminal system in Manhattan.[38]

If the report had been more successful, it might have had a negative effect on plans for a vehicular tunnel. But it had two fatal flaws. First, it proposed a terminal plan that was not the highest and best use of Manhattan land and therefore actually harmed the argument for introduction of rail lines into Manhattan. As the *New York Times* pointed out in an editorial on February 2, 1918, there was new antagonism between those who had thought that all freight cars ought to be run over tracks into Manhattan and those who thought that they should be excluded in the interest of higher uses of the land for manufacturing, retail trade, and residential development. The problem with the first idea was obvious. "The several railways ending at the Hudson's banks in New Jersey can never have sufficient wagon tracks or switching facilities in Manhattan," the paper noted. "They can have both in New Jersey, and their motors or private motors can deliver freight in Manhattan better than lighters can do it."[39]

The second fatal flaw of the commission's report was that in recommending creation of a Terminal Improvement Commission, it ignored the existence of the New York, New Jersey Port and Harbor Development Commission and intruded on the territory of the two state bridge and tunnel commissions. The report's recommendations, and the bill introduced in the New York legislature to carry out those recommendations, were therefore strongly opposed by influential organizations such as the Public Service Commission and the Citizens Union.[40]

Despite the report's shortcomings, within days of Van Benschoten's initial announcement of its findings, many other voices echoed its conclusions, calling for an end to the delays that had stifled progress toward construction of an all-weather trans-Hudson transportation link. As the *New York Times* reported on January 27, 1918, "The coal famine, due in a large measure to the icebound condition of the harbor, and the lack of terminal facilities, has given new impetus to the plan to either tunnel or bridge the Hudson River as a means of furnishing a quick connection between the New York and New Jersey shore for all kinds of vehicular traffic."[41]

It was Van Benschoten, however, who perhaps best surmised the feelings of New Yorkers during the coal famine. "As the people . . . watch the ineffectual attempts to transport the coal to the needy and shivering city," he stated, "the mistakes and failures of the past are surely brought home, and the decision must be reached that never again, if it can possibly be prevented, must such a condition exist."[42]

The general realization that the conditions experienced during the winter of 1917–1918 should not be repeated added strength to the position of the New York, New Jersey Port and Harbor Development Commission when it made its preliminary report to the state legislatures in February 1918, requesting an appropriation from each state of $400,000 to conduct an in-depth investigation of freight-transportation problems in the port.[43] The legislatures reacted favorably, each appropriating $100,000 in 1918 and the same amount in 1919. When those funds proved inadequate, an additional $25,000 was appropriated by each legislature in 1920.[44]

The basic assumptions guiding the Development Commission in its investigation were dictated by the ICC findings. The ICC, with its eye on past practice and existing conditions, did not appreciate the future importance and possibilities of the motor truck as a means of hauling freight. The Development Commission compounded that shortsightedness by considering only the effect that the proposed vehicular tunnel might have on ferry-passenger traffic, without regard to the effect that such a tunnel might have on freight movement. When it made its final report in 1920, it indicated clearly that it never considered the vehicular tunnel of any importance as a link in the system of freight transportation. "The Commission has not felt it to be within its province to attempt to forecast the effect the vehicular tunnel or tunnels will have on the Port problem as a whole," the report stated. "The four highway bridges between Manhattan and Long Island have rendered the East River ferries obsolete, but the first bridge did not. Experience with rapid transit facilities in New York has shown that the traffic increases as rapidly as the systems expand, leaving as large a burden as ever on the older facilities; and the same may prove true with the vehicular traffic of the tunnels and ferries."[45]

The shortsightedness with which the Development Commission began its work was also revealed in another portion of its final report, where it briefly considered the contributions to freight movement that could be made by a bridge over the Hudson River carrying both passenger rail and motor-truck traffic. The commission, the report stated, "feels that it has analyzed the highest development of motor-truck service . . . and has found the service uneconomical." In reference to its finding that the existing ferries were adequate to the needs of motor vehicles, the commission found "that the existing ferries, without either bridge or vehicular tunnel, could handle the entire Manhattan tonnage."[46]

This was not true. But the commissioners believed, from the start of their work, that it was true. With the direction of the commission's future investigation dictated by a cursory examination of past practice and by a lack of vision regarding the future, it was incapable of analyzing a mode of transportation that others saw as of increasing importance, not just for movement of people but for movement of freight.

The ICC decision, the report of the Commission on West Side Improvements, and the report of the New York, New Jersey Port and Harbor Development Commission, all officially released early in 1918, illuminated the issues faced by public decision-makers regarding the best way of solving the transportation-related problems of the Port. A comprehensive solution, however, would not soon be found. In the meantime, the work of the bridge and tunnel commissioners in achievement of a partial remedy via construction of a vehicular tunnel continued.

In February 1918, the New Jersey legislature created the New Jersey Interstate Bridge and Tunnel Commission, thus legislating out of existence the Hudson River Bridge and Tunnel Commission and the Delaware River Commission. The new bridge and tunnel commission would now be responsible for both projects. The legislature also appropriated $10,000 for expenses.

On March 12, the New York State Bridge and Tunnel Commission made its eighth annual report to the New York legislature. The report briefly summarized the history of the New York and New Jersey bridge and tunnel commissions and then outlined the need for a tunnel, stating, "A vast amount of food, fuel and merchandise could be expeditiously transported by truck through the tunnel direct from New Jersey freight yards to wholesale and retail merchants throughout New York City, unhampered by weather conditions such as extreme cold, ice interference or fog." In an attempt to remind the legislators of the unmet needs of babies and children during the recent weather crisis, the report then provided an example of one benefit to be realized, pointing out that a tunnel "would revolutionize the bringing of milk into the city every night, and the milk wagons would no longer be dependent on transportation by ferries."[47]

In a section devoted to the need for prompt action, the report stated, "Such a situation as was witnessed this winter seemed impossible a week before it occurred, and yet it is now a matter of history and may easily occur again, both in respect to fuel and food." The report also noted that the city "needed the fateful combination of circumstances to bring out in startling relief, and in a way that must never be forgotten, the imperative need for a remedy and the danger of further delay." That remedy, of course, was a vehicular tunnel and not a freight-rail tunnel. In citing the report of the Commission on West Side Improvement, the eighth annual report agreed that railroad tunnels for delivery of freight would be useful, but such tunnels were part of a comprehensive plan of terminal development and freight distribution that would take years and vast sums of money to implement. A vehicular tunnel, however, would "give immediate relief from entire reliance upon harbor and river transportation, and can be built now without the disbursement of a great sum of money." In addition, what the commission was suggesting was "nothing but the construction of a highway between two states, and as such is not involved with the problem of freight terminals."[48]

In regard to financing, the report noted that the state of New Jersey had proposed to fund its share by a direct tax over a period of four years and that New York would "undoubtedly provide its share of the cost by a bond issue." The shares to be borne by the states would be lessened by participation of the federal government, and, the report stated, "we are informed that persons in authority would favor legislation by Congress to that end."[49]

The eight annual report of the New York commission was favorably received by the New York legislature, which subsequently reauthorized its bridge and tunnel commission and appropriated about $3,000 for expenses. It could not, however, pass any appropriation exceeding $1 million without a referendum, so it was unable to fund its share of the initial construction costs.

On June 14, 1918, the two state bridge and tunnel commissions held a joint meeting at 115 Broadway in New York and elected Thomas McCarter chairman of the "joint commission." The joint commission also appointed Edward A. Byrne, chief engineer of the New York City Department of Plant and Structures, as chief engineer of the commission, leaving Goethals in the position of "consulting engineer." Byrne had already exerted his influence on Goethals by convincing him that the design needed to be changed to accommodate three lanes of traffic in each direction instead of just two.[50]

The commissioners agreed that the first step toward obtaining federal funds for half the cost of the tunnel would be the introduction in Congress of a bill to that effect. The New York members of the commission agreed to ask Senator William M. Calder, a Republican from Brooklyn, New York, to introduce the appropriate legislation in the Senate. The New Jersey members agreed to approach Congressman John J. Eagan, a Democrat from

New Jersey, to simultaneously introduce a bill in the House of Representatives. The bills were successfully filed on June 28. At this point, New Jersey had already made provision for its share of the construction cost, but the New York legislature adjourned before passing the law necessary to provide funding from that state.[51]

In the summer of 1918, as the bridge and tunnel commissioners made plans to appear before the Senate Committee on Interstate Commerce, a change occurred on the New Jersey commission that was to have profound consequences. In July, McCarter attacked the Board of Public Utility Commissions of the State of New Jersey, calling the members "political horse thieves." The board had just refused to grant a request of the Public Services Railway Company (a Public Service Corporation subsidiary) for a rate that McCarter believed enough to allow it to pay a reasonable dividend. That refusal, according to McCarter, practically ruined the credit of the company. But McCarter's attack was not just an insult to the board; it was an insult to Governor Edge, who had appointed the board. Although McCarter later apologized for his name-calling, he would not retract his criticism of the board's decision. This apparently led to a falling-out between McCarter and Edge, which resulted in McCarter's resignation from the bridge and tunnel commission.[52]

In August 1918, Edge appointed Thomas Albeus Adams of Montclair, New Jersey, to replace McCarter as a member of the commission. As the *New York Evening Post* later claimed, this appointment was made "only after considerable debate."[53] Just who was involved in that debate is unknown, but it surely involved the New Jersey commission's Chairman Noyes, who knew Adams well. Noyes was vice president of the New York division of Swift and Company, which had business relations with companies owned by Adams, and both were members of the New Jersey State Chamber of Commerce.

Adams had previously been appointed to the tricounty Hudson River Bridge and Tunnel Commission in 1917 and was shortly thereafter elected vice chairman. He soon became one of the strongest advocates of the tunnel, and according to the *New York Times*, he "was a factor in reviving interest in the plan when it was moribund."[54] He had good reason to advance the tunnel plan, as he stood to benefit personally from the project.

Although Adams had lived in Montclair, New Jersey, since about 1912, he was a New Yorker. He was born in Troupsburgh, Steuben County, New York, on September 5, 1864, and educated in the public and private schools of New York State. After obtaining his secondary education, he taught school and studied law before taking a job with Swift and Company. He was appointed general manager for New York City and vicinity and held that position for about ten years.[55]

In 1895, Adams and three of his friends formed the New York Credit Men's Association, which he later incorporated as the National Credit Men's Association. His interest in finance also included presidency of the Gansevoort National Bank of New York City, located at 354 West Fourteenth Street, which was set up in 1889 to meet the needs of companies in the nearby meatpacking district. Adams served as president of the bank from 1898 to 1905, when he sold out his controlling interest for more than twice what he originally paid in January 1901.[56] He was also a director of the Mercantile National Bank when that organization merged with the National Broadway Bank and the Seventh National Bank in April 1903.[57]

In 1898, Adams and his brother Robert A. Adams formed Adams Brothers Company, a New York City–based wholesale meat and provision distributing firm with depots in the principal cities of the eastern states. This company was very profitable, and in 1905 the brothers sold out to Swift and Company. Adams was also president and director of the Manhattan Refrigerating Company. Originally incorporated in 1894, this firm operated a power plant and several cold-storage warehouses on the block surrounded by Gansevoort, Horatio, Washington, and West Streets at the north end of the meatpacking district. The company installed a system of underground pipes through which it supplied refrigeration to buildings throughout the district.[58] In 1907, Adams and his brother, along with T. W. Taliaferro, acquired the controlling interest in two affiliated firms, the Union Terminal Cold Storage Company in Jersey City and the Kings County Refrigerating Company in Brooklyn. The plant of the Jersey City firm was located on the block bordered by Provost Street on the east, Henderson Street on the west, Twelfth Street on the south, and Thirteenth Street on the north. This placed it squarely in the path of the most likely alignment of the vehicular tunnel.

In January 1918, the Union Terminal Cold Storage Company held 1,997 of the 2,000 outstanding shares of the Provost Realty Company. Adams was president and director of both companies. Adams held one share of the Provost Realty Company individually, as did his brother Robert A. Adams. In June 1918, Provost Realty acquired a twenty-five-by-one-hundred-foot section of the block on Twelfth Street between Grove and Henderson. This block was just west of the block occupied by the Union Terminal Cold Storage Company.[59]

As president of the Chelsea Association of Merchants and Manufacturers, a New York organization, Adams sent out ten thousand promotional circulars in 1918, printed at his own expense. Titled "Reasons for the Construction of the Vehicular Tunnel between New York and New Jersey," these brochures set forth the advantages of the project. He also initiated a letter-writing campaign, having his associates in the New Jersey and New York business community send letters in support of the vehicular tunnel to legislators in both states.

After bills authorizing the tunnel were introduced in Albany and Trenton, and after preliminary studies had all but fixed the termini, Natalie Jarvis, a widow, sold the majority of a block bounded by Twelfth and Thirteenth Streets and by Provost and Barnum Streets, just across Provost Street from the Union Terminal Cold Storage Company, to Katherine M. Wallace, mother-in-law of Adams. One small part of the block on Twelfth Street (a twenty-five-by-one-hundred-foot parcel), surrounded by the property formally owned by Jarvis, was owned by heirs of Ellen Geary. In a deal made on November 8, 1918, Provost Realty also acquired this tract. On November 28, 1918, Provost Realty acquired two additional tracts of land on Thirteenth Street, just east of Provost Street. This is roughly where the Jersey City exit of the tunnel would have to be located, given the general alignment that had already been determined.

Throughout 1918, as Adams made plans to profit from the project, he confidently asserted to members of the New Jersey and New York commissions that he had sufficient influence with both state legislatures to have the enabling and appropriation acts for construction of the tunnel passed. Apparently, this was no idle boast. As the *New York Evening Post* later stated of Adams, "As chairman of James R. Nugent's machine in Essex County it was expected that he would be able to exert some influence, but the pressure that he brought to bear seemed to be all out of proportion to the influence that might have been his as a Democratic county chairman of a machine that was in disfavor with the party power."[60]

By the end of 1918, however, Adams's fellow bridge and tunnel commissioners focused their attention less on the pending actions of state legislatures than on the tunnel appropriation bill then pending in Congress. Given that approximately half the nation's freight moved through the Port of New York, it seemed reasonable to the commissioners that the nation should participate in funding the tunnel as a vital improvement in the port's transportation system. And with federal participation, the state legislatures would find it much easier to appropriate their shares of the construction cost. The key to unlock the federal cash box would be a successful argument that the facility was an improvement of national and not just state importance. As the time for making the case before Congress drew near, the commissioners, their engineers, and their legal counsel prepared to make that argument, knowing that the fate of the enterprise was at stake.

4

The Wedding Ring

It gets down to this, boys. This is just the wedding ring. What happens afterwards we don't have to talk about tonight.
—New York Governor Al Smith, to New Jersey Governor Walter Edge
and legislative leaders from New Jersey and New York, regarding
their agreement to work together for bistate cooperation in
construction of a vehicular tunnel, January 29, 1919

Lieut. Governor Walker . . . declared [that] the vehicular tunnel was the ring which symbolized union of the States of New York and New Jersey, which, he said, were interdependent on each other.
—*New York Times*, October 13, 1920

ON DECEMBER 12, 1918, the Senate Interstate Commerce Committee met in room 410 of the Senate Office Building in Washington, D.C., to consider a bill to appropriate $6 million for the Hudson River vehicular tunnel. The commissioners present were Emanuel W. Bloomingdale (then chairman of the joint commission, following McCarter's resignation), George R. Dyer, and A. J. Shamberg from the New York commission; and Weller H. Noyes, Franklin Murphy, and Palmer Campbell from the New Jersey commission. Goethals and Byrne attended as the joint commission's consulting engineer and chief engineer, and Paul Windels attended as counsel to the joint commission. Also there to give support were Murray Hulbert, director of the Port of New York; Jere E. Tamblyn, representing the Chamber of Commerce of the State of New York; and Richard E. Meade of the National Highway Association. Edge had intended to be there but missed a train in Chicago while returning from a vacation and could not make it in time. Al Smith, newly elected as governor of New York, chose not to be there because his election had been contested, and he did not feel it appropriate to attend as governor elect.

Committee Chairman Ellison D. Smith of South Carolina called the hearing to order and then asked Calder to state the purpose of the bill. Everyone in the room already knew that the bill provided for construction of the tunnel by the federal government, in connection with the two states. What the committee really wanted to hear was the justification for federal participation. As Calder acknowledged,

> This, of course, is a most unusual measure. So far as I know, no legislation of this character has ever been enacted by Congress; but we believe that unusual conditions exist at the port of New York, which are no longer local, but national in bearing. From that port were shipped seventy percent of our soldiers who went abroad in this recent war emergency, and eighty percent of the things the soldiers needed. The city of New York is about to spend some $50,000,000 in the further improvement of the port, and we believe in connection with the transportation of troops and of mails and of power and other things, that the federal government can do no better than give its money to so laudable an enterprise.[1]

After then apologizing for and explaining the absence of Governors Edge and Smith, Calder called Goethals as his first witness. Making it clear that the tunnel would be for pedestrians and vehicles only, Goethals stated that the ferries operating in the vicinity of the proposed tunnel termini carried approximately three million vehicles a year. The tunnel that he proposed would have an ultimate capacity of thirty-five to thirty-six million vehicles per year. The tunnel would carry both the freight of the railroads terminating in northern New Jersey and their passengers, who would be transferred to buses. He also reminded his audience, "The question of the operation of trains across from the Jersey side to get into connection with the New York shore is under consideration by a joint port commission."[2]

In response to the question of Senator Frank Kellogg of Minnesota, who asked why Congress should build a tunnel for New York City when it turned down requests to build bridges across the Mississippi River and other rivers, Goethals said that he was not prepared to answer. "That, I understand, is for this committee to decide."

Bloomingdale did a better job of addressing Kellogg's concerns, pointing out that over 50 percent of the nation's foreign trade annually passed through the Port of New York, and "the merchant in Oklahoma or in Wisconsin, and, in fact, all over the country, who has goods shipped to and from New York, is ultimately paying the additional expense due to the delay involved by the tying up of motor trucks along the water front in New York City."[3] The project, therefore, was of national and not local concern. Resolutions in favor of the tunnel from the Chamber of Commerce of the State of New York, the Motor Truck Association of Manhattan, the Motor Truck Association of America, and the Board of Trade of the City of Newark, New Jersey, were then entered by Bloomingdale as evidence that support for the project was not just local. Speaking after Bloomingdale, Noyes stated the unanimous belief of the New Jersey commissioners that the federal government should support all interstate transportation projects across navigable rivers, as such projects addressed national as well as local problems.

Windels was the last to speak. He informed the senators that in order for the two states to have an enforceable contract for construction of the tunnel, federal ratification was necessary. When asked by Senator Oscar W. Underwood of Alabama if such legislation was necessary even if there was no appropriation of funds by Congress, Windels replied that that was the case. He also addressed questions raised by Kellogg regarding why Congress should not also fund the construction of terminal facilities to solve the problems of the Port of New York, saying, "We look upon this tunnel, Senator, as nothing but an extension of the Lincoln Highway—nothing but an interstate road. Congress has precedent, because they have appropriated large sums of money for the construction of roads, and this is the most vital 9,000 feet of road in the country."[4]

The Lincoln Highway, of which Windels spoke, was the nation's first coast-to-coast highway, linking Times Square in New York to Lincoln Park in San Francisco. Carl G. Fisher, an auto-parts manufacturer from Indianapolis, Indiana, best known for developing the Indianapolis Motor Speedway, first proposed the highway in 1911. When initiated in 1913, and for many years thereafter, the privately supported highway was little more than a designation attached to a suggested line of travel, which motorists could follow on their way across the country. Drivers beginning their journey from New York took the Courtland Street ferry across the Hudson River to Jersey City, where they had to negotiate narrow and highly congested streets on their way out of town. It was not until passage of the Federal Aid Road Act in 1916 that federal financial assistance was available to the states for improvement of the Lincoln Highway and other interstate routes over which the U.S. mail would be transported. That aid, however, was designated primarily for construction and improvement of rural roads within states and not bridges and tunnels between states.

Kellogg responded to Windels's statement about the importance of a highway across the Hudson River by saying that he had no doubt about it but added that the Lincoln Highway "crosses many streams between the East and the West, and the question is, to what extent the Federal Government is going to build bridges and tunnels." Windels then asked, "It comes down to a question of degree, doesn't it?" To which Kellogg replied, "Yes; a question of degree."[5]

With that statement, the basic issue was framed in a way that meant defeat for the proposal. Congress had already determined, through passage of the Federal Aid Road Act, that federal participation in road building was in the nation's interest. But the senators could not afford to establish a precedent that would oblige them to help fund expensive bridge and tunnel projects across the country, even if they connected two states by an interstate highway. The committee therefore voted the bill down that afternoon, with Senator Kellogg leading the argument against passage.[6] Although tunnel proponents continued to state in public that they had not given up hope of eventually securing congressional financial backing, the issue had been settled. From that point forward, tunnel proponents would have to rely solely on the New York and New Jersey state legislatures for financing.

The New Jersey legislature had indicated its commitment to fund its half of the project, but there were still powerful parochial interests in New York opposed to bistate cooperation of any sort, and legislators in that state were slow to respond. Walter Edge knew that in order to spur the New York legislature to action, he needed first to convince New York Governor-Elect Smith to support the tunnel project. But Smith, a Democrat, might be expected to oppose the creation of a port authority and construction of a vehicular tunnel, since Tammany Hall, the corrupt Democratic political organization that controlled much of what occurred in New York City, was opposed to both. Smith, however, had demonstrated a penchant throughout his political career for acting independently. Moreover, his life history and experiences provided a firm foundation for his appreciation of the problems of freight distribution in Manhattan.

Smith was born December 30, 1873, in a house that would be underneath the Manhattan approach to the Brooklyn Bridge when that structure was completed more than nine and a half years later. His Irish Catholic father owned a small trucking business, when "trucking" was conducted with horse-drawn wagons. Al dropped out of school at the age of fourteen, just a year after his father's death, and went to work as a "truck chaser," delivering messages to teamsters along the East River waterfront. In 1892, he found higher-paying employment as an assistant bookkeeper in the Fulton Fish Market, but after a few years he moved on to a better job as shipping clerk for a company in Brooklyn.[7]

After securing a minor political-patronage job in the Office of the City Commissioner of Jurors in 1895, Smith won election to the New York State Assembly in 1903 as a Tammany Hall candidate. His career in the legislature included positions as chairman of the Ways and Means Committee in 1911, minority leader in 1912, speaker in 1913, and minority leader again in 1914. In 1915, Smith left the legislature after winning election as sheriff of New York County, and on January 1, 1918, he assumed new responsibilities following his election as president of the Board of Aldermen of Greater New York.

Due to the absence of Mayor John F. Hylan, Smith served as acting mayor during the worst of the coal famine in 1918. Years later, recalling the famine, he wrote, "The necessity for some development at the port of New York that would make such a thing impossible in the future impressed itself upon me. Later, when I was Governor, it furnished the principal argument for me to accept the proposal to build a vehicular tunnel under the river between New York and New Jersey."[8]

Yet despite Smith's later recollection, when he agreed to Edge's request for a meeting in Albany at the end of January 1919, his support of plans for port development, including a tunnel, were far from certain. Edge, therefore, used every opportunity available before the meeting to prod Smith in the direction he wished for him to go. One opportunity arose January 9, 1919, when the Marine Workers Union began a four-day strike that held up passenger and freight movement in New York Harbor just as completely as could the worst winter storm. The workers were demanding an eight-hour workday and an increase in pay from forty-two cents to fifty cents an hour. With sixteen thousand union members off the job, almost all ferry traffic and most barge and lighter traffic came to a halt.[9]

On January 16, 1919, Edge released the text of a letter to Smith, which stated, "The recent strike has caused the business men of New York to lose more money than the tunnel would cost to construct." Edge could not conceive, he wrote, that "two enterprising and progressive communities like New York and New Jersey" could continue to allow such events to occur, "with their attendant results." Directly challenging Smith to action, Edge then reminded Smith that New Jersey had already passed the necessary legislation to fund its half of the construction cost and was waiting on New York to do the same. "You have a commission thoroughly interested in the work," the letter concluded, "and, of course, the resources to easily provide for the improvement; but New Jersey can now only await your action on the matter."[10]

On the evening of January 29, 1919, Edge, Smith, Julius Henry Cohen, Eugenius H. Outerbridge (one of the New York commissioners of the New York, New Jersey Port and Harbor Development Commission and president of the Chamber of Commerce of the State of New York), and legislative leaders from both states came together in Smith's quarters in the Executive Mansion. Just steps away from the mansion, the Capitol building was fully lighted as the New York legislature met in a late-night session to discuss whether the state would join in ratification of the Eighteenth Amendment to the United States Constitution. The amendment, already ratified on January 16 by the necessary number of states to make it law, would prohibit the manufacture, sale, or transportation of alcohol, beginning one year after ratification.

Cohen's intention in asking Edge to set up the meeting was to reach agreement on establishment of the Port Authority. The Port Authority was an idea developed by Cohen, who envisioned an entity having broad powers to control development within a "Port of New York District" that would include an area encompassing more than fifty communities in New Jersey and New York. Under the Compact Clause of the United States Constitution, the Port Authority would be created through revision of an existing treaty between the two states enacted in 1834.[11]

Edge, on the other hand, was more interested in discussing the tunnel project, and that subject took up most of the evening. As the debate dragged on, Cohen turned to Outerbridge and said, "It seems to me they are discussing the matter of one trouser button and buttonhole to save us from disgrace, while we are here to discuss an order for a whole suit of clothes."[12] Smith, however, was in no hurry to stick his neck out to promote a project unless he was certain that it had a chance of success. He was aware of widespread support for the tunnel. As just one example, earlier in the day he received a letter from the city Merchant's Association urging an immediate appropriation of $1 million to fund initial construction. But there was opposition as well, and Smith wanted to make certain that everyone in the room was fully committed to a workable agreement. Yet the debate dragged on, with no agreement in sight.

Suddenly, the phone on Smith's desk rang. He picked up the receiver, listened for a moment, and then hung up. The call had been from one of Smith's assistants, next door in the Capitol building, who had been moni-

toring debate on the Eighteenth Amendment. The Senate had voted to rat-
ify. Smith told his quests what had happened and said, "Let's have a drink.
Mine's beer, what's yours?" With that, all the men toasted New York State's
agreement to the "prohibition" amendment. The meeting then returned to
the subject at hand, without much progress until Smith, acknowledging
basic agreement on principles, if not on details, finally said, "It gets down
to this boys. This is just the wedding ring. What happens afterwards we
don't have to talk about tonight."[13]

Cohen and Edge left the meeting assuming that Smith would support
both a compact for creation of the Port Authority and a compact between
the states for construction of the tunnel. But Smith was politically astute
enough to know that he could not take the lead in advocating either proposal. His strategy would be to present himself as a supporter of whatever
"the people" wanted. His first step in gauging support for the Port Authority was to urge passage, on February 5, of a legislative resolution appointing
the governor, lieutenant governor, and various senators and assemblymen
as a commission to work with a similar body from the New Jersey legislature to write a treaty for port development. On February 15, the two commissions met in New York at the Chamber of Commerce to work out a
draft. At that meeting, Smith made it known that he was opposed to any
treaty that would give the proposed Port Authority absolute power to dictate harbor development. Similar sentiments were expressed by others.
New subcommittees composed of six members from each state were then
created to work out the details with the Port Development Commission,
but, as is so often the case, disagreement on details proved to be the undo-
ing of the plan. It was not long before opposition in New York began to
unravel what had been so tightly woven by Cohen.

On March 1, the Interstate Conference on Harbor Development met
at the Chamber of Commerce offices at 65 Liberty Street. The tentative
draft of the new port treaty had already undergone considerable change to
restrict the powers of the Port Authority, but the New York legislators informed their counterparts from New Jersey that the administration of New
York City (meaning Tammany Hall) objected to the plan. As Smith told
the representatives from across the river, New York State did not feel safe in
proceeding further until the plan had been presented to, and approved by,
the people. "We must ask your patience," he said, "for we have problems to
wrestle with here in New York State that you have not." He then further explained that the New Yorkers were still "intensely interested" in the proposition, but if they asked for more time, it is only so that the treaty could be

"very carefully considered by the people of New York, and any difficulties ironed out."[14]

As later became obvious to all, a treaty for creation of a Port Authority with real power would not be worked out, at least not for many years. Realization of that fact may have actually benefited the tunnel plan, however, as the need for some type of immediate action to solve the problems of the port was soon reasserted. On March 4, another strike of marine workers tied up about 90 percent of the harbor traffic. This time the strike lasted through the end of the month. Politicians and business people, with yet another reminder before them of the vulnerability of the city to the existing freight-transportation system, began to call more vociferously for a tunnel. Thaddeus C. Sweet, speaker of the New Jersey Assembly, stated on March 5, "Completion of this tunnel will afford an opportunity for transporting the necessities of life regardless of harbor strikes." He also expressed his confidence that Goethals, who had retired from the army at the first of the month, would be able to accept bids for construction that fell within the $12 million estimated for the project.[15]

Across the river, the Board of Trade and Transportation decisively rejected the proposed port treaty at a meeting on March 12. Opposition to the plan was so strong that a motion to ask Cohen to come before the board to explain those portions of the plan to which members objected was voted down. For the majority of those present, their minds were made up. As they saw it, the treaty would give representatives from New Jersey the power to veto improvement projects originating in New York which were exclusively located on the New York side of the harbor. Yet the board adopted a resolution favoring the tunnel.[16]

Smith thereupon began to focus his energies on the tunnel plan. He first asked the newly created New York State Reconstruction Commission to look into the tunnel project and to make recommendations for further steps to take. This organization, ostensibly designed to reorganize and reform the state government, was also created to win support from progressive Republicans who were suspicious of Smith's Tammany Hall connections.[17] The commission found that the methods proposed by Goethals and O'Rourke to build the tunnel were so new and untried that it chose to submit three questions to five well-known engineers: Can the type of tunnel proposed be constructed? If it is constructed, would it be useable? and Are the estimates of cost approximately correct? The engineers selected to pass judgment on the plans were Amos Schaeffer, consulting engineer of the borough of Manhattan; William B. Parsons, former chief engineer of the city's Rapid Transit Commission; George W. Fuller, a New York City consulting engineer known for his expertise in water management and sanitary engineering; George F. Kunz, a mineralogist and member of the New York Academy of Sciences; and Paul G. Brown, formerly the engineer in charge of constructing the two-and-two-thirds-mile-long Lake View Tunnel under Lake Michigan. The inclusion of Fuller and Brown on the team of consultants was particularly important because the top layer of the Hudson River bottom between New York and New Jersey was silt created by dumping raw sewage into the river. With their expertise, these men would be able to address the risks of corrosion posed by this material on the tunnel's concrete outer ring.

In the middle of March, the consultants released their report, which affirmed that a vehicular tunnel between the suggested termini was greatly needed and should be built as soon as possible. The plan proposed by Goethals and O'Rourke, however, had problems. Not only would the shield used in boring the tunnel be of unprecedented size, but there was also no precedent for the use of concrete blocks. These two factors alone made it very difficult to accurately estimate the ultimate cost. In addition, it was unknown whether the concrete blocks would be sufficiently waterproof under hydrostatic pressure. In contrast, the cost and time to complete two more-conventional cast-iron tunnels of smaller diameter, each providing for two lanes of traffic, could be accurately estimated. "It is recommended, therefore," the report concluded, "that any plans, specifications, and estimates for the construction of a vehicular tunnel or tunnels under the Hudson River between New York and New Jersey should be approved by a committee of competent engineers before a contract is awarded for construction."[18] Smith, of course, could not disagree with his own commission, so the creation of a board of consulting engineers was henceforth an indivisible part of the tunnel plan.

In the middle of March 1919, the New York Senate advanced the tunnel funding bill, which had already passed the Assembly, to a third and final reading, after an amendment had been made prohibiting the Public Service Corporation of New Jersey from using the tunnel to run its street cars into Manhattan. The bill was opposed by the Tammany delegation, but Smith stayed in the background. On March 29, he announced that his mind was entirely open on the issue of whether the tunnel should be funded. In order to measure public support, he would hold a Saturday meeting at City Hall on April 5, at which time representatives of every interested organization in the city could call on him and express their views on the tunnel bill. He would not, however, discuss the proposed port treaty.[19]

On the day the meeting took place, about 150 citizens, representing civic associations, real estate interests, taxpayers, and other business organizations showed up to voice either their opposition or support. Queens Borough President Maurice Connelly, who claimed that New Jersey already had better freight facilities than New York did and lower tax rates, opposed a vehicular tunnel. He stated that if the proposal was to build a rail-freight tunnel, he would be in favor of it, but, he said, "until New York City is put on the same footing with New Jersey I shall oppose it."[20] Representatives of the Long Island Real Estate Exchange and the Queens Chamber of Commerce also opposed the plan, on the grounds that industry would be attracted to New Jersey, which would profit more by the tunnel than would New York.

Most of those in attendance favored the project, with the Merchants Association, the United Real Estate Owners Association, the Broadway Association, the Greater New York Taxpayers' Association, the National Highway Traffic Association, and the Bronx Board of Trade all indicating their support. As just one example, George W. Olvaney, representing the Central Mercantile Association, expressed his belief that the cost of the tunnel would be less than that of the present harbor strike.[21] John O'Rourke attended the meeting, and he informed those present that the tunnel could be built for $11.5 million (half a million less than his previous estimate) and that he would post a bond for the full amount. He also pointed out the advantages of one forty-two-foot tunnel but said that he would build twin tubes if that plan was adopted. Smith then thanked those attending and declared the meeting adjourned.[22]

Smith had heard enough. The people had spoken. Six days later, he signed the tunnel bill and sent a telegram to Edge, saying, "I take pleasure in saying that by signing the appropriation on behalf of New York for the construction of the tunnel, the Empire State reaches the hand of friendship across the Hudson to greet her sister State and neighbor—New Jersey."[23] Continuing this spirit of cooperation expressed in Smith's telegram was now in the hands of the two state bridge and tunnel commissions, which would have to work together closely to bring the project about. That spirit was immediately tested, however, as each commission approached the task from a different perspective.

The New York State Bridge and Tunnel Commission members for 1919 were Chairman George R. Dyer, Vice Chairman Emanuel W. Bloomingdale, Alexander J. Shamberg, McDougall Hawkes, Frank M. Williams, and Grover A. Whalen. Paul Windels served as legal counsel, and Morris M. Frohlich served as secretary. Bloomingdale, Shamberg, and Hawkes had served on every New York State bridge and tunnel commission since being appointed to the original commission in 1906. Dyer was first appointed when the number of commissioners expanded in 1907, and he had been chairman of all the New York commissions since 1913. Generally referred to as "General" Dyer, he had been commander of U.S. forces on the Mexican border in 1916 and commander of the New York National Guard during the First World War. Bloomingdale was brother to Joseph B. and Lyman G. Bloomingdale, who developed Bloomingdale Brothers into one of New York City's great dry-goods and retail businesses. He represented the Retail Dry Goods Association of New York for many years, served as a presidential elector in 1900, and was involved in a number of civic organizations. Shamberg was a cattle dealer and head of the exporting firm J. Shamberg and Son. His business required him to travel back and forth to Jersey City by ferry, and, like many others, he sought an alternative that would allow people to drive across (or under) the river in their own automobiles. More than any other person, he was responsible for creation of the first commission in 1906. Before the appointment of Williams, Hawkes was the only engineer on the commission. He had served as city commissioner of docks from 1902 to 1903 and did much to improve the Hudson River waterfront. New York State Engineer Williams and New York City Commissioner for Plants and Structures Whalen both served as ex-officio members, in accordance with the enabling legislation. Whalen, former secretary to New York City Mayor John Hylan, still had the most illustrious phases of his career ahead of him. He was known primarily as the city's "official greeter" until 1928, when he became police commissioner.

The members of the New Jersey Interstate Bridge and Tunnel Commission for 1919 were Chairman Weller H. Noyes, Vice Chairman Samuel T. French, Palmer Campbell, Richard T. Collings, Thomas J. S. Barlow, Daniel F. Hendrickson, Theodore Boettger, and T. Albeus Adams. Emerson L. Richards served as legal counsel, and E. Morgan Barradale served as secretary. In accordance with legislation passed in 1918, this group was responsible for both the Hudson River tunnel and the Delaware River bridge projects. Four of the commissioners—French, Barlow, Hendrickson, and Collings—were from south New Jersey and were interested primarily in the Delaware River bridge. The other members had interests directly related to trade across the Hudson River. Noyes had first been appointed to the New Jersey Interstate Bridge Commission in 1910, and he had been chairman since 1913. He was vice president of the New York division of Swift and Company, the country's largest meat-packing firm. He was also a prominent member of the New Jersey Chamber of Commerce and president of the Tenafly Trust Company. Campbell, general manager of the Hoboken Land and Improvement Company, was one of the wealthiest people in Hoboken, New Jersey, and one of the founders and the first president of the Hoboken Board of Trade. He was also vice president of the Hudson County Park Commission and had been identified with the business, civic, and social life of the county for several decades. Like Noyes, he was an active member of the New Jersey Chamber of Commerce. Boettger was president of the United Piece Dye Works of Lodi, New Jersey, the largest dyeing company in the country. Also active in philanthropic work, he was a member of many leading social clubs of New Jersey and New York.

Adams was a holdover from the previous commission. In February 1919, while the tunnel bill was under consideration by the New York State Assembly, he organized a letter-writing campaign in an attempt to impress on the legislators the importance of the tunnel to the commercial interests of New York City. Many New York and Jersey City businesses whose success depended on reliable and cost-effective shipment of products within the port area participated in the campaign, including Gimbel Brothers (department store), Perkins-Goodwin Company (paper-mill supplies), Kings County Refrigerating Company (Adams's cold-storage warehouse company), Carscallen & Cassidy (commission merchants and dealers in hay and grain), George F. Hinrichs, Inc. (dressed poultry, eggs, etc.), Peter J. Carey Print (another company in which Adams had an interest), Columbia Machine Works and Malleable Iron Company (electric railway supplies), Worthington Pump and Machinery Corporation, National Biscuit Company, House of A. Silz (poultry, game, and meat supplies), *Life* magazine (dependent on reliable supplies of newsprint), and Northwest Electric Equipment Company. Also participating were several organizations representing commercial interests and businesses with an interest in the consulting contracts that might be generated by the project, such as the Chelsea Association of Merchants and Manufacturers, Walter Kidde & Company (engineers and constructors), Electrical Club of New York, Gansevoort Market Business Men's Association, and Hickey Contracting Company (contracting engineers). The Automobile Club of America also sent a letter advocating prompt action on the tunnel legislation.

The majority of letters sent to the legislators repeated certain arguments in favor of the tunnel, including the inefficiency and cost of the present ferry-and-barge-based transportation system, the disruption of operations caused by the coal famine of 1917–1918, the New York Harbor boatmen's strike of 1918–1919, and the relative cost-effectiveness of the tunnel. The overall message conveyed is well summarized in the letter to Assembly member Clarence C. Smith (Republican, Saratoga County) from the Electrical Club of New York, which stated, "freight and express and vehicle communication and transport is so poor across the river that it forms a very great obstacle to the expansion of our business amongst the people of this district and this obstacle would be removed by the proposed tunnel."[24] Each New York legislator also received a copy of a promotional pamphlet, "Reasons for the Construction of the Vehicular Tunnel between New York and New Jersey," produced by Adams as president of the Chelsea Association of Merchants and Manufacturers. Adams printed ten thousand of these brochures, in at least three editions, at his own expense.[25]

Adams's efforts on behalf of the tunnel project were know to the other commissioners when they first met as a joint body to begin their work in the spring of 1919, but there is no evidence to indicate that they knew of his real estate deals.

The first task before the commissioners was to appoint a board of consulting engineers and to select a chief engineer. The board's duties would be to meet regularly, to advise the commissions and the chief engineer on the general type and specific design of tunnel to be built, and to follow the progress of the work during contract letting and construction. The final decision on what type of tunnel to build, and by what method, would rest with the commissioners. The chief engineer's duties would be to organize a staff, to gather data, to prepare cost estimates, to prepare plans and specifications for construction bidding, and then to see that the work was properly carried out. The chief engineer would make a recommendation as to type, size, and location of the tunnel, and if there was any division of opinion between the board and the chief engineer, the commissions would have the final word.

Goethals, with his recommendation for the type of tunnel already under fire and given the limitations imposed on the authority of the chief engineer to manage without interference, decided that he no longer wanted to head the project. In April 1919, he refused to attend an important conference to discuss the tunnel. In a letter to his son explaining his decision, he wrote, "I am no longer interested, since their mathematics as applicable to the concrete block is all wrong, and this was my only interest or concern. I wouldn't take the job now of any consideration." He also sent a letter to the New Jersey Speaker of the General Assembly, stating that his only interest in the future would be to defend his plan.[26] At this point, it appears that Goethals had not refused to serve on the board of consulting engineers, if only because he saw that group as the best forum for proper consideration of his recommendation.

After learning of Goethals's decision, the commissioners wasted no time in beginning their search for engineers to serve on a consulting board and for a chief engineer. A general invitation was extended to "engineers of recognized standing, particularly those who had been identified with the great accomplishments in tunnel building in the vicinity of New York City and elsewhere," to attend a series of conferences where the project was discussed. Many candidates were interviewed, with an emphasis placed on selection of an engineer who would guide the work.[27]

James Hollis Wells, a New Jersey bridge and tunnel commissioner from 1912 to 1917, was briefly considered for the position of chief engineer. But his professional experience fell more along the lines of architectural engineering. He was best known as designer of the twenty-two-story office buildings of the Hudson Terminal, built in Lower Manhattan from 1907 to 1909 for the Hudson and Manhattan Railroad Company.[28]

A more obvious candidate was J. Vipond Davies, now the sole head of his consulting firm following the retirement of Jacobs. Born in South Wales in 1862, Davies attended Wesleyan College and the University of London before beginning his engineering career working for coal-mine owners and steel manufacturers. He came to the United States in 1889 and, like Jacobs, went to work on railroad projects in New York. Later, he served as chief engineer for the New York and New England Railroad and for the Long Island Railroad, before teaming up with Jacobs on the East River Gas Company Tunnel.[29]

On April 25, when the two state commissions met for the first time as a joint commission, they asked Goethals and Davies to present their plans. Goethals adhered to the design that he had already recommended for a single tunnel with an outer shell made of interlocking concrete blocks. The tunnel would have a double road deck, with one deck to be used by eastbound traffic and the other for westbound traffic. Davies advocated a twin-tube tunnel with cast-iron outer shells and two lines of traffic in each tube. After considerable debate, it was decided that Davies would serve with Goethals on the board of consulting engineers.[30]

Sometime in late April or early May 1919, the commissions began to consider a chief engineer candidate from Brooklyn who was known for his work building subway tunnels under the East River for the New York City Public Service Commission. Clifford M. Holland, division engineer for the commission since 1916, had been a tunnel engineer his entire career, and he had a reputation for technical and organizational excellence. He was well-known in his profession as a member of the board of directors of the American Society of Civil Engineers, a member of the American Association of Engineers, and a past president of the Harvard Engineering Society. The man selected as chief engineer would have to stand between Goethals and Davies on occasion and withstand various political pressures throughout the course of the project. Some of the commissioners were concerned that Holland, at thirty-six, was simply too young for the job.

On the afternoon of Thursday, May 29, while the selection process for project engineers moved toward its conclusion, the Down Town Club in Jersey City hosted a dinner in honor of T. Albeus Adams. Hailed as the man who "put the vehicular tunnel over," he was given a silver loving cup by the Hudson County Team Owners Association. Thomas Stewart, president of the association, told the audience of the tireless energy displayed by the honoree and of the considerable sums of money he spent during his year-and-a-half lobbying effort in Albany. Rising to speak while the group cheered, Adams deprecated the praise showered on him but admitted that he had done "a little hustling" to move the legislation along.[31]

The project that Adams had worked so long and hard to advance was stalled, however. It was rumored that there had been considerable friction among the tunnel commissioners regarding the appointment of engineers, and there were public rumblings of discontent concerning the joint commission's closed-door meetings. Several months had passed since enactment of the enabling legislation, yet no engineers had been chosen, and no plan or design had been definitively agreed on.

On June 3, 1919, the commissioners met and finally resolved their differences concerning appointments to the board of consulting engineers. In addition to Davies, the members would be William J. Wilgus, Henry W. Hodge, John A. Bensel, and William Burr. Goethals was out. Goethals, displaying a behavior pattern evident throughout his life, had declined to work as part of a group. Without complete freedom to make decisions and authority to carry out those decisions, he could not function. He also could not accept the fact that the commissioners were willing to consider a tunnel design other than the one he had recommended. To him, they were amateurs with no ability to pass judgment on engineering design, and he balked at the notion that they would have the right to reject or modify his plans. From this point forward, he continued to defend his plan, but only as an outsider.[32]

The engineers selected were capable men. Hodge was a partner in Boller, Hodge and Baird, the consulting engineering firm that had studied the possibilities of a bridge over the Hudson River for the bridge and tunnel commissions in 1913. He had also been a New York State Public Service Commissioner. His tenure on the board was brief, however. He died December 21, 1919. Burr, former head of the engineering school at Columbia University, had served on the board of consulting engineers that developed a tunnel plan for the New Jersey PSC in 1917. Bensel had a range of experience in large construction projects. He was the former New York City commissioner of docks, and as president of the city's board of water supply, he oversaw construction of the Catskill Aqueduct System. He was also a former state engineer for New York State, in charge of canal construction. During the First World War, he worked to improve harbors around the country.

Wilgus, soon to be elected chairman of the board, was destined to play a role in early project development and construction greater than that of all other consulting engineers and second only to that of the chief engineer. He was born in Buffalo, New York, on November 20, 1865. His father was a foreman at the New York Central Railroad's Carroll Street Freight House. After graduating from high school, Wilgus took a job as apprentice draughtsman to a local civil engineer and studied drafting through a Cornell University correspondence course. In 1885, he left home to work as a rodman on a railroad surveying crew and took on progressively more responsible positions with a number of railroads before returning to Buffalo in 1895 as chief engineer of the Buffalo Terminal Railway Company. In 1897, he became resident engineer of the New York Central and Hudson River Railroad, and by 1903, he advanced to vice president of construction. He oversaw the design and construction of New York City's elegant Grand Central Terminal during its early stage of development and retired from the railroad in 1907.

As an advisory engineer for the Michigan Central Railroad in 1905, Wilgus developed a method of building a two-track tunnel in watertight sections, on dry land; the sections were then sunk into trenches dredged from the bottom of the Detroit River and connected underwater. His work on the Detroit River Tunnel won him the Telford Gold Medal from the Institute of Civil Engineering of Great Britain.

In 1908, as a consultant to the Amsterdam Corporation (where he was also a partner), Wilgus submitted a plan to the New York State Public Service Commission for freight transportation in the greater New York City region. The plan featured an underground electric freight railway in Manhattan using small cars running through approximately sixty miles of small-bore tunnels, about eleven feet beneath the city streets. The system would also extend under the Hudson River to New Jersey, where there would be another 34.5 miles of track. Later, his plan for improving freight distribution in the port area expanded to include two concentric rail beltways: an inner beltway that would use bridges and tunnels to move freight around the port area without the need for watercraft, and an outer beltway that would allow trains to bypass the port.[33]

In 1909, Wilgus also began developing an idea for an interstate metropolitan district that would have a geographical coverage of nineteen hundred square miles and powers similar to those later proposed for the Port of New York Authority. Although he was ahead of his time, and thus unsuccessful in winning enough support for his plan to work, his advocacy of a joint federal and state investigation of the needs of the port was an important factor in the eventual creation of the Port of New York Authority.[34]

Throughout Wilgus's tenure on the tunnel board of consulting engineers, he continued to advocate his plan for a regional freight-rail distribution system and fought to keep the railroads from taking over the Hudson River vehicular tunnel for their own purposes. The vehicular tunnel, in Wilgus's view, would not be a solution to the problem of freight distribution in the Port of New York but would serve merely as a first, and partial, step in execution of a regional plan for transportation improvement.[35]

In addition to announcing the appointment of Wilgus and the other consulting engineers on June 3, the commissioners announced their pick of a man the New York Times called "the youngest chief tunnel engineer in the United States, and probably in all the world."[36] They could not have made a better choice. Clifford Holland seemed destined for the job. A handsome man of average height and build, with thin hair and a receding hair line, Holland wore thin-framed glasses that helped give his face an open, inviting appearance. In photographs, he always seemed to have a slight, Mona Lisa–like smile. Born in Somerset, Massachusetts, on March 13, 1883, Holland was an only child. He attended public schools in Michigan and Massachusetts before graduating from the Cambridge Latin School in Cambridge, Massachusetts, in 1902. He entered Harvard University that year, and despite having to earn part of his college expenses by teaching evening school, waiting tables in the college dining hall, and reading gas meters, he graduated in three years with a bachelor of arts degree in 1905 and followed that with a bachelor of science in civil engineering in 1906. Before graduation, Holland had passed civil-service examinations and been appointed assistant engineer for the Rapid Transit Railroad Commission of New York City (predecessor to the PSC). The day after receiving his engineering degree, he left Cambridge for New York, telling a friend, "I'm going into tunnel work, and I'm going to put a lot more into it than I'll ever be paid for."[37]

His first project was the Battery–Joralemon Street Subway Tunnel, under construction since 1903. Although he only worked on the project from July 1906 to its completion in 1908, he gained valuable experience in the shield method of compressed-air tunneling and on lining the inside of the cast-iron shell with concrete, grade corrections, construction of the ventilation system, and, by observation of his superiors, project management.[38] He supplemented his practical experience by reading every article that he could find on the subject of subaqueous tunnel construction.

From 1908 to 1914, there being no new subway-construction projects under way, Holland completed the construction records of the Battery-Joralemon Tunnel and then took charge of a portion of the Fourth Avenue subway project in Brooklyn. He was later assigned to study the optimal design to be recommended for future subway tunnels. His specific task was to gather data as to the dimensions, depths, and construction conditions of all similar tunnels that had been built in the United States and Europe, compile the results, and make a comparative analysis of their engineering features.

On November 5, 1908, he married a woman he had known since they were both students at the Cambridge Latin School. Anna Coolidge Davenport was born in Boston, and she attended Radcliff College while Holland was at Harvard. Their first child, Anna Hesketh, was born December 8, 1909, followed by Clarissa Coolidge on June 5, 1911.

In 1914, the PSC awarded contracts for the Old Slip–Clark Street Tunnel and the Whitehall–Montague Street Tunnel, and assistant engineer Holland was put in charge of both construction projects. When contracts were later awarded for the Willoughby–Montague Street, Fourteenth Street, and Sixtieth Street Tunnels, he was placed in charge of those projects as well. As evidence of his skill and the high esteem in which he was held by the laborers who worked on his projects, a foreman on the Sixtieth Street Tunnel said to a visiting reporter, while jerking his thumb in Holland's direction, "That bird could come down here blindfolded in the dark and tell us if we were going wrong."[39]

In 1916, the PSC promoted Holland to division engineer, the position he still held when selected to take on the chief engineer's duties for the Hudson River tunnel. Another child was born to the Hollands on July 13, 1916: their third daughter, Benita Davenport. While he was working on the East River tunnels, Holland told his wife that someday the Hudson River would be conquered in the same manner, and he would like to be the one to do the work. Often he spoke of the details of how it would be done. As Anna recalled years later, "It was in his power to visualize a big engineering task and to see it as a whole in his mind." When he was offered the job, he was "overjoyed."[40]

Each member of the board of consulting engineers was to be paid a salary of $10,000 a year, as was Holland, even though his duties were far more extensive and his responsibility greater. Holland told the two commission chairmen, Dyer and Noyes, that the chief engineer's salary should be $15,000 a year. In response, they said that he was being offered a lesser amount because he had never been a chief engineer before and was "untried" in that position. Holland accepted, with the stipulation that he be given a free hand in choosing his assistants and that they be paid adequately.[41]

Holland's preliminary selections for staff (and their annual salaries) were Jesse B. Snow, principal assistant engineer ($5,400); Milton H. Freeman, resident engineer ($4,200); Ole Singstad, designing engineer ($4,200); Ronald M. Beck, assistant engineer ($3,000); Rheata M. Latti, stenographer ($1,200); and Howard H. Stiegwewald, clerk ($1,200). The inclusion of Snow, Freeman, and Singstad was of particular importance, because of their prior association with Holland as tunnel engineers for the PSC. Without their agreement to devote many years of their lives to the project, Holland would not have taken the job.

Snow, fifty-one, was born on Nantucket Island and was a descendant of early Cape Cod settlers who had entered the shipping business in Massachusetts. He graduated from Union College with a civil engineering degree in 1889 and spent the early years of his career on a variety of sewer, waterworks, street railway, railroad, and bridge projects in New York, Wisconsin, and Canada. In 1902, he was appointed chief engineer of the Jersey

City Transit Company, and from 1905 to 1910, he worked on the Pennsylvania Railroad Tunnel under the Hudson River. With the completion of that project, Snow left New York to work on reconstruction of the Hauser Lake Dam near Helena, Montana, and on a large hydroelectric project in Montreal, Canada. Snow joined the PSC as a resident engineer in 1914 and worked with Holland on five subway-tunnel projects under the East River.

Freeman, forty-eight, was born in Crary Mills, St. Lawrence County, New York. After graduating from the Potsdam Normal School in 1895, he taught in the New York State public schools for two years. From September 1897 to June 1899, he was principal of Heuvelton Union Free School in Heuvelton, New York, but left to enter the University of Michigan to study mathematics and physics, with the intention of teaching these subjects at the secondary level. After beginning his college studies, he developed an interest in engineering and graduated in 1903 with a bachelor of science in civil engineering.

Freeman, like so many other young engineers before him, gained his first practical experience as a railroad survey crew member. In September 1905, he began work on the East River tunnels of the Pennsylvania Railroad, first as an inspector of the shield used for driving the tunnels and then as assistant engineer. From May 1909 to September 1914, he worked on tunnel projects for the New York City Board of Water Supply, before returning to rail-transportation-related work as resident engineer (the engineer in charge on-site) on the Old Slip–Clark Street subway tunnel. From 1916 to 1918, Freeman performed similar work as resident engineer in charge of the river section of the Sixtieth Street subway tunnel, and from 1918 to his appointment on the Hudson River vehicular tunnel, he was resident engineer in charge of the Manhattan shaft of the Fourteenth Street subway project. Holland depended heavily on Freeman to ensure that everything went according to plan on a daily basis.[42]

Singstad, thirty-seven, was born in Norway and immigrated to the United States in 1905 after receiving his degree in civil engineering from the Polytechnic Institute of Trondheim. He worked on a variety of railroad projects, including the Hudson and Manhattan Railroad tunnels, before taking a job in 1910 with the PSC. During his seven-year tenure with that agency, he designed several of the subway tunnels on which Holland worked. When Holland asked him to join the Hudson River tunnel project, Singstad was employed by Barclay, Parsons and Klapp, designing a rapid-transit system for Philadelphia.[43]

When Holland offered Singstad the job, the Norwegian hesitated. The challenges of ventilation would be unprecedented, perhaps insurmountable. While thinking it over, Singstad took a ride on the Pennsylvania Railroad ferry across the Hudson River. As the boat approached the dock at Desbrosses Street, Singstad looked back across the Hudson River toward Jersey City and asked himself how the problem of ventilation could be met. With confidence that he could find a solution, he accepted Holland's offer.[44]

Holland filled the most important staff positions with men familiar to him through their prior association on New York subway-tunnel projects. This also meant that these positions were filled by men from New York. The New Jersey commissioners had no great problem with these senior appointments, but there were twenty-nine positions initially authorized for appointment, including assistant designing engineer, chief of survey party, mechanical engineer, rodman, draftsman, leveler, and page. With the highest-paid positions going to New Yorkers, it soon became obvious to the New Jersey commissioners that these positions represented a total of $79,000 a year in salaries, while just $3,000 a year in salaries had gone to people from New Jersey.

In a sign of problems to come, New Jersey commissioner Noyes informed his counterparts from New York at a joint meeting on Thursday afternoon, June 10, that he expected a fifty-fifty distribution of appointments and that New Jersey had just as many competent people to serve as draftsmen, rodmen, and clerks as did New York. "We have splendid material in New Jersey," he stated, "just as you have in New York. We wish it understood that hereafter New Jersey will expect to be treated with as much consideration as is New York in the matter of appointments."[45]

This demand generated an acrimonious debate among the commissioners, which was duly reported by the *Jersey Journal*. Published accounts of friction between the New Yorkers and the New Jerseyans proved embarrassing to both and threatened to tarnish the appearance of unity that the commissioners wished to display. At a meeting of the New Jersey commission on June 17, Commissioner Campbell introduced a resolution, inspired by Noyes, to censure the *Journal* for airing dirty laundry. After discussion, the resolution was withdrawn, but Noyes got some of what he wanted. Consideration would be given thereafter to a potential appointee's place of residence.

With the board appointed and the engineering staff ready to begin work, one important item remained to be addressed. A contract between the states that would govern their agreement to build the tunnel had to be signed. The New York commissioners took the lead in preparing a draft, which was submitted to the New Jersey commission on May 20. The contract included a clause that would protect either state in the event of a default by the other. This was necessary because each state had funded only the initial $1 million necessary to begin the work. If either state failed to pass additional appropriations in the future, the other state could build the tunnel by itself and charge tolls at both termini until the full cost was repaid.

The commissioners signed the contract in their joint office at 115 Broadway on September 23, 1919. The "wedding ring" that Smith, Edge, and legislative leaders from both states had figuratively slipped on their collective fingers nine months before was, at last, backed by a legal union. The greatest political barrier to progress toward construction of a vehicular tunnel, that of interstate rivalry and mistrust, had, at least for the time being, been overcome.

Now it was time to agree on exactly how the tunnel should be designed and built. To achieve consensus, the tunnel commissioners relied heavily on the recommendations of their staff engineers, as confirmed by the Board of Consulting Engineers. Those engineers had been chosen for their technical expertise, wealth of experience, and professional objectivity. But they soon found their design task complicated by the political agendas of the commissioners and by the personal interests of their peers in the profession, many of whom hoped to profit by adoption of a type of tunnel or method of construction in which they had a financial interest. Goethals had foretold this problem in a letter to his son written the previous March: "I have been fighting for my tunnel, but whether it goes through or not, I don't know. The Legislature will pass the bill all right, I guess; then will come up the question of type, and then the real trouble will begin."[46]

5

A Controversy Acute and Personal

A controversy that was growing to be acute and personal has been quieted by the commission charged with building the Hudson River vehicular tunnel, through its decision against the much-discussed 42-ft. concrete-block tunnel plan. With regard to the technical questions involved, the decision is to be commended because it is favorable to progress, and in this particular enterprise nothing is so badly needed as prompt and rapid progress.

— *Engineering News-Record*, March 18, 1920

HOLLAND AND HIS assistants began their design study by reviewing all the existing plans and reports that had been made in regard to the tunnel, including those produced by Jacobs and Davies, the New Jersey PSC, and Goethals. They then addressed the most fundamental consideration, that of location. The previous February, Adams succeeded in having a clause introduced in the New York enabling legislation requiring that the Manhattan terminus of the tunnel be in the "vicinity" of Canal Street. This was in accordance with the suggestions of everyone who had studied the issue since at least 1913, since Canal Street was at the approximate center of the downtown Manhattan vehicular traffic area, as determined by the already established location of ferry docks and terminal facilities. But Adams pushed for the clause to ensure that the tunnel location would maximize the value of his real estate holdings.[1]

A tunnel terminus near Canal Street would provide easy access to the wholesale business district of Manhattan, as well as to the large number of warehouses located on the lower west side. This location would also give the most direct connection to the three downtown East River vehicular crossings, the Brooklyn, Manhattan, and Williamsburg Bridges. The Manhattan Bridge was directly in line with Canal Street and had the largest capacity for vehicular traffic. It would, therefore, carry the majority of tunnel traffic to and from Long Island.

The New Jersey legislation did not have a similar clause fixing the terminus on the Jersey City side, but the logical point would be opposite Canal Street in order to keep the tunnel length as short as possible, thus keeping the cost down. Canal Street was also located opposite the main traffic arteries in Jersey City leading to Hoboken, Newark, and other industrial centers in New Jersey. The exact location of the entrances and exits in both cities still needed to be worked out, so that traffic coming from and flowing into the tunnel would meld with that of surrounding streets, while also keeping land-acquisition costs to a minimum.

Traffic capacity was the second major design consideration to be addressed. Holland's main concerns were the volume and type of traffic that would use the tunnel; the capacity of one, two, or three traffic lanes in each direction; the most economical size of the tunnel in relation to the traffic demand; and the limitation placed on capacity by the ability of the New York and Jersey City street systems to handle congestion.

Up-to-date counts of vehicles using the ferries were made during the summer and early fall of 1919 so that accurate projections of future traffic could be calculated, thus informing decisions regarding the tunnel's design capacity and potential revenue. The engineers used historical trends to predict the level of traffic demand from 1924, the anticipated date of tunnel completion, to 1943, the date at which the largest tunnel proposed (the Goethals design, with three lanes in each direction) would reach maximum capacity for a combination of horse-drawn and motor-vehicular traffic. Their figures showed that overall demand would grow to such an extent that traffic on the ferries would continue to increase even with construction of a tunnel. There was no question, therefore, that demand would be sufficient to generate enough revenue to reimburse the states for the cost of tunnel construction and operation within a reasonable period of time. The only issue to be decided was the number of traffic lanes to be constructed in each direction. The engineers concluded that one lane would not carry enough traffic to generate sufficient revenue, while three lanes would carry more than the city street systems could handle. Therefore, two traffic lanes in each direction was the most logical choice.

The next issue to consider was how to provide adequate ventilation at a reasonable cost. The studies conducted by the New Jersey PSC in 1916 were useful and provided some basis for preliminary designs. But clearly, new studies would have to be done to determine the amount and composition of the exhaust gases created by motor vehicles, the dilution with fresh air necessary to render these gases harmless, and the method and equipment necessary for economical ventilation. Since this type of investigation was outside the area of expertise of Holland and his staff, they arranged to have the United States Bureau of Mines conduct a series of tests to obtain the needed data.

At the direction of the Board of Consulting Engineers, Holland also began consideration of the type and method of construction. What was the physical nature of the riverbed, and how might its composition determine what methodology to use? Should the tunnel be made of steel, cast iron, or concrete? Should the tunnel segments be constructed on shore, towed out to the river, and sunk into predredged trenches, or should a shield be driven through the riverbed, with the tunnel's outer ring built in pieces as the shield advanced?

On January 6, 1920, Holland submitted his first official report (dated December 31, 1919) to the commissions. The engineers had made a detailed study of eleven different plans. The one recommended, developed by Holland and his staff, was for twin shield-driven, cast-iron tubes with exterior dimensions of twenty-nine feet, containing twenty-foot-wide roadways to accommodate two lanes of traffic in each direction. The cast-iron shells would be lined with concrete.

The entrance to the tunnel in Jersey City would be at Provost and Twelfth Streets, with an entrance plaza extending another block to Henderson Street, occupying the entire block bounded by Henderson, Provost, Eleventh, and Twelfth Streets. The New Jersey exit would be two blocks north of the entrance at Provost and Fourteenth Streets. This plaza, smaller than that for the entrance because no toll booths would be needed, would require widening Fourteenth Street between Provost and Henderson Streets by taking one hundred feet from the properties on the north side.

The entrance to the tunnel in New York would be at Broome Street between Varick and Hudson Streets, with a plaza occupying most of the area between Broome Street and Watts Street. The New York exit would be on the south side of Canal Street at Varick Street.

The projected cost of the tunnel, considering the salvage value of real estate and easements and an allowance for contingencies, was $28,669,000. The estimated time for completion was three and a half years.

After reviewing the chief engineer's report, four members of the Board of Consulting Engineers wrote an endorsement, which included a statement that the views of Henry Hodge, who had died on December 19, 1919, "are known to have been in accord with those expressed in this report." The omission of Byrne's name on the letter was explained by his absence "on account of illness."[2] Byrne had, in fact, missed all Board meetings from December 9, 1919, to February 9, 1920, due to illness. But as the other Board members were about to find out, if he had been in attendance, he would not have signed the letter of endorsement because of concerns about tunnel capacity. His absence also meant that he missed out on a crucial discussion regarding Holland's estimated cost for the project.

Wilgus believed that Holland's estimate was too low. He subsequently sent a letter to the commissions, dated January 21, 1920, in which he provided a detailed explanation for his charge, citing three main areas of concern. He began with Holland's estimate of the salvage value of real estate and easements, which the chief engineer had used to reduce net cost by $1.2 million. According to Wilgus, Holland had "no real basis for his salvage figure." He was also concerned that Holland's estimate of 5 percent for contingencies ($1,365,000) was too low by about half, because of the unprecedented scope of the project, and took exception to Holland's decision to exclude as a cost item interest charges that would accrue during construction. He argued that the total cost of the project should be placed at $31,354,000 if interest during construction was excluded, or $35,116,000 if that item was included, with the understanding that this cost would be reduced by the salvage value of real estate.[3]

After considerable debate, the other Board members sided with Holland, even though they apparently recognized that Wilgus had a valid argument. As it turned out, Wilgus was justified in his concerns, and the cost estimate given in the chief engineer's report would have been more honest, albeit still too low, had the other Board members and the commissioners listened to Wilgus.

The chief engineer's report and the letter of endorsement from the Board were adopted by both commissions on January 27, 1920. Each commission then made its own report to its legislature. The New York commissioners, in order to receive public input on their decision, announced that two public hearings would be held in the United Engineering Society Building at 29 West Thirty-Ninth Street: one on Saturday evening, February 21, to discuss traffic circulation, and the other on Monday evening, February 23, to discuss the general plan. Byrne let it be known that he would present a dissenting report at one or both public hearings.[4]

In a letter sent to the Board on February 16, 1920, Byrne stated that he did not agree with the estimated capacity of the tunnel or with the proposed location of the New York entrance and exit, and he was unconvinced that a cast-iron lining was the best option. He believed that Holland's estimated capacity of thirty-two hundred vehicles per hour for the two-way, twenty-foot-wide roadways was too high and that to equal or somewhat exceed that capacity, it would be necessary to build three lanes in each direction, each twenty-four feet, six inches wide, omitting the two-foot-six-inch sidewalk. The estimated increased cost of the larger tunnels required by his proposed changes would be $6 million.

The other Board members felt that Holland's estimate of capacity was correct, but even if it had not been, the extra width of roadway proposed by Byrne was too narrow for three lanes of traffic. The sidewalk had to be included because of provisions of the New York enabling legislation. They also saw that Byrne's plan would result in such serious congestion at the exits that it would nullify the added capacity provided by three lanes of traffic. Like Byrne, some of the New Jersey commissioners found it difficult to accept the chief engineer's recommended plan. Goethals, after all, had been *their* state engineer. All of them except Adams, however, agreed to endorse the report and abandon the Goethals design.

On the morning of Tuesday, February 18, 1920, Goethals retuned to New York from a trip to Washington, D.C., and was shown newspapers which quoted Holland's criticisms of the Goethals plan. He immediately fired off a letter to the bridge and tunnel commissioners asking for copies of the commissioners' reports. He also included a list of five items he wanted to see regarding the data and analyses backing up Holland's claims.

Holland's response was that it would be impossible to provide everything that Goethals had asked for, because it would take about three weeks to assemble and he was working with a limited staff. "General Goethals," he offered, "is at liberty to examine for himself all of the data in the office or it will be placed at the disposal of any persons whom he may designate to gather the information for him. He has just as much right to see the records of the commission as any other private citizen."[5] Goethals, seeing this response as an insult, refused to go down to the engineer's office and instead prepared to defend his plan at the public hearings.

A few days before the first scheduled hearing, Byrne went to a meeting of the New Jersey commissioners and informed them that he still opposed the plan recommended by Holland and intended to submit a memorandum asking for revisions. Adams, who by this point had become a proponent of the Goethals-O'Rourke design, asked Byrne to return on February 20 and formally present his complaints to the commission, with the press in attendance. When the time came, however, the other commissioners refused to show up, probably thinking that the public hearings would be the proper venue for the airing of objections. Without a quorum, there was no meeting. Adams, thoroughly embarrassed, explained his actions to reporters by stating that he was opposed to the adoption of any plan until the public had its say.[6]

The first public hearing, on February 21, 1920, provided little excitement, as the somewhat dry topic of traffic demand was presented and discussed. This was not an unimportant aspect of the chief engineer's report. As stated by the *Engineering News-Record*, the leading professional engineering publication of the time, "Outranking all the discussion and analysis of tunnel types, the traffic studies on which the design is predicated have great engineering importance. They demonstrate with almost startling clearness the urgent necessity of providing better means for crossing the Hudson than the present ferries."[7] The public hearings, however, were attended mainly by engineers and contractors who wanted their design selected, and to them the traffic studies were irrelevant.

The second hearing on February 23, devoted to a discussion of the overall scheme of the project, was livelier. Holland made a lantern-slide presentation, which included a discussion of the various plans that had been subject to review, with an emphasis on the reasons behind rejection of the Goethals-O'Rourke plan: because there was no adequate method of tying the individual blocks of the ring together, the tunnel ring could not withstand the permanent stresses, let alone the stresses during construction; the concrete-blocks would not stand up to the corrosive effects of sewage-laced Hudson River silt, which was too unstable to hold the unanchored tunnel in place; the twenty-two-foot-six-inch width of the roadway was not wide enough for three lanes of traffic, as planned, and could only carry two lanes; the ventilation ducts were too small to adequately ventilate the tunnel; the construction cost was too low and the projected three-year time to complete too optimistic; and the methodology of construction was generally untried and unsafe. After Holland concluded his presentation, Byrne's memorandum of dissent was accepted for the record. A number of invited speakers were then heard, and the floor opened for general discussion.

Goethals, who had been listening quietly as Holland attacked his plan, got up to speak. In a voice shaking with emotion, he said, "I am deeply interested in this matter because I advocated the concrete tunnel plan recommended by the commissions last year. It was on the basis of this plan that they asked appropriations from their Legislatures. I personally was asked to write letters to members of the Legislature about it, and now I am told that the plan is no good."[8] With a flushed face, Goethals continued, addressing the charges that the concrete blocks could not stand the differential pressures of ever-shifting Hudson River silt and might corrode in the sewage-contaminated bed of the river. "The Hudson River silt, which has been so much talked about tonight is no different from that in other rivers, the Ohio, Miami and the Columbia, as well as at the ends of the Panama Canal. I have built concrete structures in all of these, and they will be standing long after I am dead."

In regard to the difficulties and potential cost overruns that might be expected from use of a new construction method, Goethals stated, "So far as 'untried methods and devices' are concerned, if a man has the courage of his convictions, if we are sound in our principles of construction, and if we are willing to take the responsibility, the concrete block tunnel can be put through successfully at a price much less than has been proposed by the present plan." This brought a round of applause from the engineers and contractors who wished to see the Holland plan defeated, either because they did not believe that it was the best plan or because of self-interest.

O'Rourke then took the floor. He declared he could construct the tunnel for "one-half the cost and in one-half the time" required for the plan recommended by Holland and asserted that the chief engineer was wrong in his description of the Hudson River silt. It was "firm and dry" and would give ample pressure on the concrete shell. O'Rourke then announced that he would turn over his patents for concrete-block construction and grouting to the New York Chamber of Commerce for any compensation that the chamber thought reasonable, if those patents were used for the tunnel. In response, George H. Snyder of Jacobs and Davies (the firm still retained Jacobs's name, despite his death the previous September) pronounced the O'Rourke plan "unsafe and an unwarranted expenditure of public funds." He said the concrete tunnel would collapse.

Other engineers and contractors also presented their plans. T. Kennard Thompson offered to build a trench tunnel which would be "bigger, safer and cheaper" than the Holland type. This plan was one of those given due consideration by Holland and his staff and was found to be unsuitable because of the conditions of the riverbed, which precluded a trench tunnel. J. C. Meem of the Frederick L. Cranford Company, who had worked on the Battery-Joralemon Tunnel, proposed three different tunnel designs. Although Holland had met with his former associate on several occasions to discuss these plans, he found them without merit. Meem, no doubt to the surprise of many people in attendance, said a steel-shell tunnel with concrete-block lining could be built in thirty days!

The hearing was capped, however, by Paul G. Brown, a member of a United Engineering Societies committee that had studied the tunnel problem for Governor Smith. Brown said that not a single engineer familiar with tunnel building in the East and Hudson Rivers favored the Goethals plan and added that the Holland report was the best he had ever seen on an engineering project.

With the two New York public hearings behind them, the New York commissioners, the Board, and the engineering staff announced their intention to put the consideration of type and methodology of construction behind them as well. All their efforts would now be directed at preparing for the advertising of bids in midsummer so that ground could be broken in early August. But with the New Jersey public hearings still ahead, the fight was far from over.

On Thursday, March 4, 1920, the third public hearing and the first in New Jersey was held in City Hall in Newark. Adams, acting as chairman of the hearing in the absence of Noyes, said in opening remarks that the commissions had not approved the recommendations in the report of the chief engineer and that approval would depend on what the public said at the public hearings. Otherwise, there would be no need for hearings. He further stated that the purpose of the meeting was to give an open opportunity for criticism of the plans of the chief engineer and to hear of any other suggestions and plans which those present might bring forward, to help the commissions reach their final conclusions.[9] This was not true. The New York commission had already sent to the New York legislature its report endorsing the Holland recommendation, and the New Jersey commission had voted to forward its positive recommendation of the Holland plan to the New Jersey legislature by the end of the month. But Adams did not allow anyone to refute his statement.

Holland then gave his well-honed lantern-slide presentation, fully explaining the recommendations of his report. As soon as he finished, Adams directed the assistant secretary of the New Jersey commission to read Byrne's minority report, in spite of the fact that both commissions had failed to take any action other than ordering it filed with their records. Wilgus asked if he could read the Board's majority response to Byrne's report, but Adams refused.

Adams then declared the hearing open to presentation of oral or written statements regarding alternative suggestions or plans. Goethals led off, saying, "The statements that have appeared in the press that concrete is not a suitable material for tunnel building is [sic] not supported by the facts." He then went on to state that concrete was a more economical structural material than cast iron and that using it would result in a savings of $10 million. He also claimed that a single tunnel was more economical than two tunnels. He concluded by arguing against following the precedent established by use of cast-iron outer rings in previous Hudson and East River tunnels, saying that followers were in the great majority, while leaders were few.[10]

O'Rourke, in a change of tactics, submitted cost estimates for a pair of thirty-one-foot-diameter concrete tunnels, which he claimed could be built at about half the cost of the cast-iron tunnels recommended by Holland. He also repeated all the claims that he had previously made regarding his willingness to release his patents, if a suitable profit was guaranteed.

Engineers and contractors who were advocates of other patented systems of construction or who had other personal interests to serve in their recommendations also spoke. St. John Clarke, a consulting engineer from New York who had been chief engineer of the Steinway Tunnel (also known as the Belmont Subway Tunnel) under the East River, spoke in favor of the use of reinforced concrete and offered to bid on an alternative plan that he claimed would save several million dollars and a year of time. J. C. Meem also claimed that he could save a year in time and proposed three different methods of construction. Practically all of those engineers who were present and independent of any connection with competing schemes spoke unqualifiedly in support of the twin-tubes plan approved by the two commissions. Two of the most prominent were Brown, who had also spoken at the second New York hearing, and Edward Wegmann, a recognized expert in dams and water-supply structures.

Around midnight, after about half the audience had left the hall, Adams finally told Wilgus that he could read the Board's majority response to Byrne's minority dissent, but Wilgus yielded the floor to Davies so that he could refute some errors and misstatements which had been made by the advocates of the precast-concrete-block tunnel.

Holland, his staff, the Board, and, most importantly, the members of both commissions were outraged at the way the hearing had been conducted. In a joint conference on Tuesday, March 9, the commissioners denounced Adams and voted to send statements to the governors of both states indicating their displeasure with his behavior. In the course of a heated discussion, Holland asserted that "General Goethals is leading a retreat" and is "thirty years behind the times" and that while the five major objections to his plan had been given in the published reports of the tunnel engineers, "at least eighty-two" defects had been found in the Goethals

plan in studies begun in July and continued for many months. Growing increasingly agitated, Holland declared that Goethals "does not know what
he's talking about" and, in reply to a request from Adams for specifics, repeated his often expressed opinion that the block-type tunnel could not
be made watertight at the joints, having "'absolutely no connection at all
between the rings." Wilgus interjected that the tunnel was in effect "a bar-
rel without hoops." Holland also said that O'Rourke was still evolving his
process, had been forced to make four changes in his method already, and
would have to make "hundreds more" if he undertook to construct the proposed tunnel.

The New York commission's Chairman Dyer then announced his desire
that Byrne resign from the consulting board. Noyes defended Byrne's right
to express a dissenting opinion but called Adams's method of conducting a
public hearing "outrageous" and promised that the next hearing, to be held
in Jersey City on March 16, would be conducted differently. It was agreed
to have the statements of the consulting board, in reply to Goethals and
others, read at that final public hearing.[11]

At a meeting of the New Jersey commission later the same day, in accordance with a recommendation adopted at the joint conference, New Jersey
commissioner Theodore Boettger moved that the Board and engineering
staff give no more consideration to any plan except the one recommended
by Holland. Boettger claimed that the engineering staff had already spent
about $20,000 of the public's money studying the Goethals-O'Rourke
plan, and enough was enough. At the suggestion of Wilgus, who believed
that there might be merit in the use of a steel lining, the motion was modified so that the Board could still consider alternative designs, except for the
concrete-block plan. Adams's was the lone negative vote. The New York
commission later agreed with this resolution.

The commissioners also resolved at their joint meeting to pay no further
attention to suggestions made by civic organizations and not to attend any
more hearings or meetings regarding type or method of tunnel construction. They did agree to attend a meeting of the National Highway Traffic
Association scheduled for Thursday evening to discuss the tunnel only in
terms of traffic demand and capacity. After that, they would listen only to
representatives of organizations invited to attend meetings in their office.[12]
If they thought that the subject of tunnel type was behind them, however,
they were wrong.

Approximately seventy-five people showed up at a meeting of the National Highway Traffic Association held at 8 p.m. on Thursday, March 11,

1920. After the meeting was called to order, a letter was read from Stuyvesant Fish—New York City financier, former president of the Illinois Central Railroad, and former business partner of O'Rourke—advocating moving platforms for conveying vehicles in the tunnels instead of having them operate under their own power. A letter from Byrne was then read, claiming that only a three-lane tunnel would provide for the projected traffic demand. Holland countered Byrne's letter by explaining that the number of lanes or width of roadway were not the factors governing capacity. Designing the entrances and exits to prevent traffic congestion was what mattered most. The floor was then opened for general discussion.[13]

The same parade of opponents who had been present at previous hearings took their turns blasting the Holland plan. This time they were joined by almost all the representatives of automobile- and truck-related organizations in attendance, who thought that the two-lane tunnels would have inadequate traffic-carrying capacity. George H. Pride, of the Heavy Haulage Company, said that Holland's estimates of future traffic were two low, and he predicted that the tunnel would be at full capacity almost on the first day of operation. And as many of the automobile representatives claimed, traffic on the existing ferries, which was the basis of the staff engineer's traffic study, was not an adequate indicator of future demand.[14]

Amos L. Schaeffer, consulting engineer to the borough president of Manhattan, defended the two-lane tunnel plan by pointing out that no single river crossing of the Hudson River could ever handle all the traffic that might wish to get from one side to the other. The tunnel capacity was absolutely dependent on the ability of the street systems at the termini to handle the traffic. It was this consideration that really dictated the optimum size of the tunnel and not whether it had two or three lanes in each direction. Wilgus joined the discussion, saying that if Byrne's recommendation for three lanes of traffic each way were followed, the result would be impossible traffic congestion at the entrances and exits.

The arguments by Holland, Schaeffer, and Wilgus fell on deaf ears. At the end of the meeting, the association adopted a resolution calling on the governors of New York and New Jersey to withdraw approval of the plan recommended by Holland and to authorize studies of a larger-capacity tunnel. The Motor Truck Association of America also sent a telegram to both governors demanding a six-lane tunnel. Although the Goethals-O'Rourke plan was not specifically mentioned by either association, it appeared that they both favored that particular design.[15]

Holland's engineering staff, with only one last public hearing before them, began to prepare contract drawings and specifications for the type of tunnel that he recommended. With sensitivity to public opinion, however, the commissioners had decided to keep the door open for alternative designs that could later be submitted in ample time for their incorporation in a request for construction bids, after action had been taken by the voters of New Jersey on the necessary bond issue in the fall.[16]

Adams's next move was an attempt to cast doubt on the motivations of the Board and engineering staff. At a joint meeting of the commissions on the afternoon of March 16, just hours before the last scheduled public hearing, Adams tried to read a detailed statement which included a claim of malfeasance by the Board. The commissioners refused to let him present his letter and instead had a secretary read it.

Dyer, who presided at the joint meeting, then asked Byrne if he and Adams were not "interested" in the Goethals type of tunnel. Byrne replied that he resented the remark and was actually inclined to prefer a steel-shell tunnel, concrete-block lined, in the interest of economy and efficiency. He had no knowledge that Adams was "interested" in any type and believed that Adams merely sought to build a tunnel as quickly, as cheaply, and as efficiently as possible. Adams also objected to Dyer's remark, claiming that his motives were just as Byrne had understood them to be. Dyer retorted that he remained convinced that Adams's interest lay with the Goethals tunnel, and Adams rejoined, "Well, you're wrong."[17]

With animosity building, the last public hearing was held later that evening in City Hall in Jersey City. The script for the drama that played out varied little from that for other hearings, with familiar actors playing familiar roles. Goethals did not attend but sent word that he was preparing a full response to criticisms of his recommended plan. In his absence, proponents and opponents of the plan recommended by Holland went at each other with gusto, and there were "sharp exchanges of opinion between the engineers of the tunnel and J. F. O'Rourke," according to one witness.[18]

The technical issues discussed were much the same as before, but with one new wrinkle. In a portent of things to come, some of those in attendance from New Jersey said that the existing road network of Jersey City was inadequate to handle the anticipated traffic, and they asked that the commissioners pay for a new Twelfth Street viaduct from the tunnel terminus to the top of Bergen Hill and for the widening of Henderson Street. This issue of which entity should pay the costs of upgrading the Jersey City street system soon became the most contentious of all.

The general feelings of the Board and engineering staff concerning the value of the public hearings were well summed up in the Board's "Tri-Monthly Report No. 2," written by Burr on April 8, 1920. "It is advisable for this Board to state that none of the hearings developed any constructive criticism whatever, or produced a single material suggestion in any way il-luminating or helpful to the engineering work entrusted to the Chief En-gineer and the Consulting Board," he wrote. "Indeed, the great majority of the engineering suggestions were either so defective or ill-considered, or even fantastic, that they were scarcely worthy of serious consideration." He also found that there had not been any serious or thorough consideration of the congestion that might be caused by traffic entering or leaving the tunnel on the New York side. "The difficulties of unmanageable congestion at Hudson and Varick Streets are sufficient, aside from every other consid-eration, to cause the rejection of any tunnel tube or tubes with a full three-way roadway in each direction," Burr concluded.[19]

With all the public hearings finally concluded, Holland and Goethals moved their battle to the pages of the *Engineering News-Record*, which published long and detailed letters from both engineers in its March and April editions. But the Board and engineering staff had the support of both commissions, Adams excluded, in their decision to ignore the Goethals-O'Rourke plan or any plan that called for three lanes of traffic in each direc-tion. Adams, however, soon had a powerful ally in his attempts to work the tunnel plan to his advantage.

Early in April, 1920, the Board learned that two of the New Jersey com-mission members whose terms had expired on February 26 were not to be reappointed. Governor Edwards had decided to replace Samuel T. French with Samuel M. Shay, a lawyer from Merchantville in Camden County. Shay was placed on the commission mainly due to his interest in the Dela-ware bridge project. Although he proved to be less amenable to coopera-tion with the New York commissioners than French had been, he was not a great opponent either.

The same could not be said of John F. Boyle, who replaced Palmer Campbell. Boyle, forty-nine, was a heavy-set man with light-blue eyes set under a heavy brow, who presented a perpetual look of discontent in pho-tographs. He was born in New York City but, while still young, moved across the Hudson River with his family when his father established a box-board company in Jersey City. After his father's death in 1905, Boyle took over the firm.[20]

In 1904, Boyle also started the Reynolds Boyle Company, manufactur-ers of wholesale paper stock, located at 500 Montgomery Avenue. Later he changed the firm name to John F. Boyle Company and made it one of the largest manufacturers in Jersey City. He expanded his business interests into real estate and finance and was a pioneer in the establishment of building-and-loan associations in Hudson County. He was a director of the Jersey City Trust Company, the Mutual Benefit Light, Heat and Power Company, and the Colonial Building and Loan Association, and he served as a member of the Board of Trade of Jersey City. Boyle's success in business led him into politics, and he became an early backer and close advisor of Jersey City Mayor Frank Hague. With Hague's help, Boyle became treasurer of the Hudson County Democratic Campaign Committee and treasurer of the Democratic State Campaign Committee.

It was Hague, no doubt, who was responsible for Boyle's appointment to the bridge and tunnel commission, using his power to dictate certain political appointments to Governor Edwards, a close friend and political ally. Hague was a Jersey City native, born there January 17, 1876. He started life in a slum locally known as "the Horseshoe." The neighborhood, through which the approaches to the tunnel would run, was populated by Irish immigrants who came to Jersey City to build the railroad yards which sprawled along the Hudson River waterfront. Hague's father was a railroad blacksmith, and after Hague's expulsion from the sixth grade for misbehavior, he also went to work in the Erie Railroad yards as a blacksmith's helper. The job lasted about two years before Hague found more lucrative work as a boxing promoter.[21]

In 1897, when Hague was twenty-one years old, he ran for and won election as constable in the Second Ward, an unpaid position that was a valuable foot in the door of local Democratic politics, which were then controlled by Robert Davis. His hard work for the party resulted in financially rewarding positions as a sheriff's deputy in 1898, precinct leader in 1901, and ward leader in 1906.

After Hague shifted allegiances from Davis to Jersey City Mayor (and Davis opponent) H. Otto Wittpenn, Hague accepted appointment as custodian of City Hall in 1908, a position that came with a good salary, control over one hundred underlings, and plenty of time to do political work. In 1911, Hague won a place on the Jersey City street and water commission. Following a shift to the commission form of government in 1913, he was elected as one of five city commissioners. By 1917, he was mayor.

Although Hague never earned an annual salary greater than $8,500, he became a wealthy man while in office by receiving kickbacks on real estate deals, gambling operations, and city employee salaries. The salary kickback, known locally as "rice pudding," was a 3 percent levy against all annual salaries and a mandatory 35 to 50 percent return on raises in salaries.[22] Of course, to be able to demand such kickbacks Hague had to control patronage, and the city payroll became bloated with unnecessary employees who were hired more for their ability to return kickbacks and votes than for any great need for the positions they held.

When the necessary legislation authorizing the tunnel project passed in 1919, New Jersey demanded its share of jobs from the project. There were not that many jobs to be doled out, however, and the ones that were available were appointed by both the New Jersey and the New York commissions, over which Hague had limited control. Hague therefore had to find another way to benefit from the tunnel. His first intention, as was gradually revealed, was to secure street improvements at the expense of the project, for which he could take credit. He also hoped to receive kickbacks from construction contracts and real estate deals connected with the tunnel project, and to do that he needed more influence with the New Jersey commission.[23] The appointment of Boyle helped give him that influence. At some point, Adams struck a deal with Boyle whereby the latter acquired thirty-five shares of stock in the Union Terminal Cold Storage Company. Boyle thus became a co-conspirator in the attempts of Hague and Adams to work the tunnel project to their advantage.

At a meeting of the New Jersey commission, held May 4, 1920, after Adams had read the Board's critical "Tri-Monthly Report No. 2," he submitted a letter that asked, "Who made the Board of Consulting Engineers the mentors and judges of the Commissions that employ them? What business of the consulting engineers is it how many public hearings are held on this subject? By whose direction do they these engineers assume to meddle in non-engineering matters with which they have no concern? And why, since they do so assume, with or without direction, do they deliberately misrepresent the results of these public hearings?" According to Adams, those hearings had shown the inadequacy of the plan proposed by the chief engineer.[24]

Adams was also upset by a reference in the report to a trip made by Burr, Davies, and Holland to Collingswood, New Jersey, on April 1 to inspect concrete which had been eroded in the sedimentation tanks for the Collingswood Sewerage Company. The trip had been made at the request of New Jersey commissioner Collings, who wanted the engineers to observe the action of sewage on some parts of the sewage-disposal plant, as being pertinent to the use of exposed concrete construction in the bed of the Hudson River, especially in the mud and silt in the vicinity of the Jersey City shore where hazardous chemicals might be found.

The trip was not without justification, as tons of untreated sewage and significant quantities of hazardous chemicals were flushed into the river every year from both states. The resulting water pollution was so bad that some of the staff engineers feared the integrity of the tunnel shell would be affected. Adams, however, claimed that the trip was "an indication of the lengths to which our engineers are going, with the consent of a majority of these Commissioners, to find means of preventing the manifest destiny of this Tunnel." Why waste time, "of which these engineers seem to be so appreciative," on such a trip? Adams asked. "We are not building a sewage disposal plant."[25]

Despite all the vitriol in Adams's letter, he did make one telling point. He noted that many engineers in favor of a six-lane tunnel had recognized that the traffic projections of the chief engineer's report failed to account for "the freight now carried by lighters that in future will go through the tunnel."[26] Although this oversight did not justify a tunnel of greater capacity, because of the limitations of the street systems at either end, it did mean that the estimates of future demand and revenue would be far off the mark.

The consideration of alternative plans, at least those involving the trench method of construction or the use of a steel tube, continued for several more months. But the design recommended by Holland and his staff became more solidified as the project moved forward.

On May 11, 1920, the New Jersey legislature passed a bill, overriding Edward's veto, to fund the Delaware River bridge and Hudson River tunnel. The law, which did not become operative until ratified by a referendum held in November, provided for a bond issue of $28 million, with a direct state tax as the means of raising the $1.6 million annual interest on the bonds. Two weeks later, on May 25, Governor Smith signed a bill appropriating $1 million as New York's share of initial construction funding.[27]

On the same day that Smith signed the New York bill, the commissioners approved the preparation of plans and specifications for construction of ventilation shafts, with two on each side of the river. Holland and his staff thereupon busied themselves with working out design details, with the expectation that a call for bids would be made by September 1.

While Holland's staff went about their work, the Board welcomed a new member to take the place of the deceased Henry Hodge. George L. Watson, of Newark, New Jersey—a consulting engineer with broad experience in mining operations and the construction of bridges, sewers, water works, and highways in the United States, Mexico, Ecuador, and Panama—was appointed June 15 and attended his first Board meeting on June 28.

At this point, despite the machinations of Adams, the Goethals-O'Rourke plan seemed to be dead. As the *Engineering News-Record* noted in its issue of June 3, 1920, "the proposed use of concrete blocks for the tunnel construction has been definitely rejected."[28] Although the plan staggered on, zombie-like, neither completely dead nor fully alive for many more months, it increasingly became irrelevant to the work of the engineers. But as one barrier to the project's advancement dissolved, another grew to take its place.

On the New Jersey side of the river, the tunnel would run underneath and emerge in the Erie Railroad yard. On July 15, 1920, Holland asked the commissions to begin formal negotiations with the railroad to acquire the necessary permanent and temporary easements and real estate. From the beginning, the New York commissioners allowed their counterparts from New Jersey to lead these negotiations. This was deemed appropriate because any land acquired on the New Jersey side would be conveyed to the New Jersey commission and not to the "joint commission," which had no authority to operate in the state of New Jersey. This abrogation of authority on the part of the New York commissioners, however, proved to be a mistake. The representatives appointed to the New Jersey negotiating committee included the chairman, Noyes, and two members from northern New Jersey, Adams and Boyle. The actions taken by these men soon threatened the entire project.

6

Political and Petty Tampering

> Political and petty tampering with an engineering enterprise of such magnitude may not end in disaster, but it will inevitably lead to a waste of the people's money.
> —*Engineering News-Record*, May 12, 1921

AS ASSISTANT TO the Chief Engineer E. R. Barradale (Holland's assistant) wrote Holland in 1923, "The history of negotiations under the Erie Agreement is very involved."[1] Barradale's understated summary gives but a hint of just how convoluted those negotiations were or how important they were to the success of the project.

On June 9, 1920, the negotiating committee made up of New Jersey tunnel commissioners held its first meeting with the Erie Railroad. By June 28, the committee had a draft agreement ready to present, which it did on August 2. The next day, after a joint conference with the New York commissioners, Boyle offered the following motion at a meeting of New Jersey commissioners, which was seconded by Boettger and declared carried: "While the proposition submitted to the Commissions by the representatives of the Erie Railroad Company is not such in its details as to be ready for immediate acceptance, the Commissions are assured that on the basis of such proposition a conclusion satisfactory to both parties will be reached, and the Committee is hereby authorized to discuss the matter with the authorities of Jersey City on that assumption."[2] The commissioners present were Adams, Boyle, Shay, Barlow, and Boettger.

The New Jersey negotiating committee then traveled across the river to meet with Mayor Hague and the Jersey City commissioners at City Hall. They left that meeting having agreed that Jersey City would vacate Eleventh Street for approximately six and a half blocks from the Erie Railroad yards to Monmouth Street and would transfer the land to the New Jersey commission, which would then transfer the vacated land to the railroad in exchange for certain grants of land and easements by the railroad in its yard east of Provost Street. Those grants and easements would allow borings and excavation for the tunnel entrances and exits and the sinking of ventilation shafts. The railroad, in turn, would be able to locate its main rail line serving the yard in the bed of Eleventh Street, something that the railroad had long desired but Jersey City had refused to do.

The railroad would also purchase and then convey to the New Jersey commission sufficient property on the south side of Twelfth Street between Provost and Monmouth Streets for the street to be widened from the existing width of 60 feet to 100 feet, except in the two blocks between Provost and Grove Streets, where Twelfth Street would be widened to 160 feet. This extra half block of width would accommodate the exit plaza and traffic leaving the plaza area. Because the exchanges would not involve tracts of land or easements with equal value, some payment of money by the tunnel commission to the railroad would also take place.[3]

The New Jersey negotiating committee also agreed that the two tunnel commissions, with each commission paying half the cost, would cover the expense of widening Henderson Street and Jersey Avenue for two blocks from Twelfth Street to Fourteenth Street and of widening Fourteenth Street to one hundred feet as far west as necessary to accommodate tunnel traffic, which would probably be to Monmouth Street.[4] This would accomplish two goals. The value of property in the vicinity of the entrance and exit plazas would be enhanced, to the personal benefit of Adams and Boyle (and possibly Hague), and Hague would be able to tell Jersey City voters that he provided needed street improvements at no cost to taxpayers.

On August 12, 1920, Adams's mother-in-law, Katherine M. Wallace, and the Provost Realty Company sold the entire tract between Twelfth and Thirteenth Streets and between Provost and Barnum Streets to Jarvis Realty Company, whose president and treasurer was John B. Wallace, son of Katherine and brother-in-law of Adams. Wallace also owned 182 shares of Adams's Union Terminal Cold Storage Company. The secretary of Jarvis Realty was Peter J. Carey, president and director of the Carey Show Print Company, of which Adams was treasurer and director.

Boyle's stake in the scheme was based partly on his ownership of stock in the Union Terminal Cold Storage Company. He was also one of the largest landowners in Jersey City and undoubtedly owned other property in the area that would increase in value as a result of the project.

If Hague had a personal interest in land that would increase in value as a result of the street widening, he was shrewd enough to keep his involvement hidden. But there were rumors. Long after the tunnel was complete, it was said that Hague used a shell company to buy real estate in the vicinity of the tunnel that was sold to the project at a nice profit.[5]

With the New York commissioners, the Board of Consulting Engineers, and the engineering staff apparently ignorant of the deal the New Jersey commissioners had made with Hague, negotiations with the Erie Railroad continued through the rest of summer and into the fall. Part of the difficulty in arriving at an agreement was that no one really knew what the vacated land of Eleventh Street was worth or what the land and easements granted by the Erie Railroad were worth. There were no recent "market value" sales of comparable properties, so the value of the properties came down to what a buyer was willing to pay and what a willing seller was willing to take, with neither party acting under duress. The Erie Railroad was not an entirely willing seller, however, because of the possibility that the New Jersey commission would have the power to condemn the land if the two parties could not agree on a value. As Holland put it in a letter to Wilgus on May 23, 1921, "Replying to your letter of May 18th in which you ask for a statement of the figures involved in the negotiations with the Erie Railroad Company, I submit the following: The value of the rights of Jersey City in the bed of 11th St. has never been definitely stated as far as I know, one man's guess being apparently as good as another's."[6]

The problem of balancing the projected costs of certain concessions to be given with the value of benefits received popped up again in November 1921, during a conference of railroad representatives and tunnel engineers. Wilgus, Holland, and Snow asked the railroad men to defend some of their figures, but "the Erie representatives frankly state[d] that they have no explanation to give."[7]

By September, the negotiations had resulted in a tentative agreement for the commissions to pay for a new pier for the railroad's use, which would be constructed over the river segment of the tunnel as it approached land. The commissioners were willing to pay for the pier because it would protect the tunnel from ships that might otherwise damage it if they were allowed to berth directly above the tunnel. But the railroad also asked for payments from the commissions that would exceed $3 million. This figure being unacceptable to the commissioners, they began negotiations in November for property and easements along the north side of the Erie yard where it abutted the Lackawanna yard, thinking that they could save some money by shifting the alignment.

While negotiations with the railroad dragged on, Holland and his staff continued to prepare plans and specifications for the first construction contract to be awarded. The engineers were under extreme pressure to finalize the plans for Contract No. 1, which would cover construction of the Manhattan land shafts, relocation of the New York Central Railroad tracks in Canal Street, relocation and reconstruction of surface and subsurface structures in Spring and Canal Streets, and construction of an engineer's field office.

Despite Holland's being authorized to add personnel to help with the work load, and benefiting from the able assistance of Freeman, Singstad, Snow, and others who were in charge of major work categories, the weight of Holland's responsibility as chief engineer increased with every new phase of the project. In recognition of the disparity between what the chief engineer was paid and what someone in his position should be paid, some of the commissioners decided that Holland deserved a raise. On June 29, 1920, Dyer recommended to the joint conference of the commissions, in executive session, that Holland's salary be increased to $15,000 per year. This salary was approved and announced to the newspapers, and congratulations poured in from many of Holland's colleagues in the profession. Shortly thereafter, Dyer asked Holland to meet him for lunch on Monday, July 26. Holland assumed that some of the commissioners had balked at the new salary, and he feared he would be asked to give up the raise.

On July 22, before the lunch could take place, Holland sent a personal letter to Wilgus, who was escaping New York City's summer heat at the Hanover Inn in Hanover, New Hampshire. "A momentous question has arisen in connection with my continuing with the Bridge & Tunnel Commissions," Holland began, "and I am writing to ask your advice, not only as Chairman of the Board but also as an Engineer of recognized standing and as a friend."[8] He had never requested the raise, he wrote Wilgus, but he did feel that it was appropriate given his duties. More important, the raise had been announced publicly, and if it were now rescinded, he would suffer embarrassment and damage to his professional reputation. For that reason, he threatened to quit if he did not receive the raise. Wilgus immediately intervened with the commissioners, the new salary was confirmed, and Holland's pride remained intact. But another challenge to Holland's status as a professional soon arose.

On August 23, Holland, the consulting engineers, and the senior members of the engineering staff—thirteen engineers in all—signed and sent a letter to the Publication Committee of the American Society of Civil Engineers (ACSE). The letter asked the committee not to allow presentation of a paper by J. A. L. (John Alexander Low) Waddell, a senior member of the society and a recognized expert in bridge engineering, at a society meeting scheduled for September 1. The tunnel engineers had received advance copies of the paper, entitled "Bridge versus Tunnel for the Proposed Hudson River Crossing at New York City," and they were very upset.

Waddell claimed that he had no objection to the proposed highway tunnel because if it proved to be unsafe after construction due to ventilation problems, it could be converted for use by subway cars, or moving platforms could be installed to carry vehicles through without use of their internal combustion engines. What irritated the tunnel engineers, however, was Waddell's statement not only that the safe ventilation of a tunnel carrying automobile traffic was an unsolved problem but that it might be unsolvable. "Figures show that such ventilation, if feasible, would be exceedingly expensive and the velocity of the passing air would be excessive." He also wrote that because carbon monoxide, like arsenic, was a cumulative poison, regular daily passage through a tunnel where the gas remained, even in minute quantities, might eventually undermine a motorist's health. Besides, he warned, there was always the chance of a blockade of traffic with the tunnel full of automobiles, and "such a blockade might result in a holocaust."[9]

The letter from Holland and the other engineers charged that Waddell's paper violated the ASCE constitution in that it advocated personal interests, and, they argued, "[It is] to an extreme degree sketchy and carelessly prepared and does not present any new matter or facts to the engineering profession; and that the subject matter of the paper is purely speculative and for the reasons thereafter stated is not in the interests of or conducive to the purposes of the Society." The paper also violated article 1, section 3, of the organization's constitution, which stated that the object of the ASCE "shall be the advancement of engineering knowledge and practice and the maintenance of a high professional standard among its members." According to the tunnel engineers, "This paper does not meet that requirement but conveys incomplete, erroneous, and thoroughly misleading statements." And lastly, the paper breached the ASCE code of ethics because it cast into question the work of those engineers employed on the tunnel.[10]

The letter did not achieve the desired result. The paper was delivered and later published as paper no. 1477 in the 1921 edition of *Transactions of the American Society of Civil Engineers*, along with short "discussions" of the paper by two New York engineers, Kennerley Bryan and Charles Evan Fowler. These men expressed fears similar to those of Waddell regarding the ventilation issue, and Fowler purported to speak for other members of his profession in writing, "There are many engineers to whom it will have to be demonstrated that a tunnel can be ventilated for automobile traffic, even at enormous cost."[11]

Judging by this letter to the ACSE, the tunnel engineers had become a little thin-skinned by this point in the project. Their reaction may have been partly due to the fact that Waddell's charge about ventilation cost and the excessive airflow speed was exactly the same as that made by Holland in regard to the Goethals-O'Rourke design. But how could any of them know what equipment would be required to ventilate a vehicular tunnel and at what velocity the airflow would have to pass through in order to ventilate? At the time, there was no precedence, no example on which to base their opinions. And that is exactly what they were—opinions—until the appropriate scientific studies were complete.

The studies being conducted by the Bureau of Mines to determine the amount and composition of exhaust gases emitted from motor vehicles were supposed to be completed by March 1 but proved to be more difficult to execute than anyone had expected. A severe winter in the Pittsburgh area made it impossible to conduct road tests, so the contract was extended to September 30, 1920. The contract called for testing passenger cars and trucks of various makes and capacities under winter and summer conditions. One hundred and one vehicles were tested, with thirty-two being passenger cars and the rest trucks from one-and-a-half-ton capacity to five tons and more. The vehicles were supplied by private individuals, corporations, and automobile dealers in Pittsburgh. In the winter tests, conducted between December 1, 1919, and February 6, 1920, the vehicles were tested while standing with engine racing; standing with engine at idle; accelerating from rest to fifteen miles an hour up a 3 percent grade; running at set speeds of three, ten, and fifteen miles an hour up a 3 percent grade; running at set speeds of three, ten, and fifteen miles an hour down a 3 percent grade; accelerating from rest to fifteen miles an hour on level grade; and running three, ten, and fifteen miles an hour on level grade. In the summer tests, conducted from March 11, 1920, to September 30, 1920, the test conditions were simplified because the winter tests had revealed that there was not a significant difference among vehicles idling, racing, or accelerating.[12]

Trucks and seven-passenger cars were tested with a full load and also with no load other than the driver, two engineers, and fifty pounds of test equipment. Five-passenger cars were tested with just a driver, two test engineers, and test equipment. On each vehicle, a test engineer disconnected the fuel line at the carburetor so that a measured amount of gasoline could be fed into the engine. The test engineer also connected an apparatus to the exhaust pipe to collect samples. For the tests involving movement, courses of a fixed length were laid out on city streets in the Pittsburgh and Highland Park area.

For the tests involving acceleration from rest to fifteen miles an hour up a grade, the vehicle was brought to a stop at the first quarter-mile mark. On the test engineer's signal, the driver would start in low gear and accelerate until the vehicle was traveling at fifteen miles an hour in high gear. The test engineer would then shout, "Stop!" and the driver would disengage the clutch and slam on the brakes. The same operation was then repeated as often as possible for a distance of one-half mile. At the time of initial acceleration, the test engineer would start his stopwatch while switching off fuel from the reserve supply and switching on the fuel from the measuring tube, while the other observer would simultaneously collect the exhaust-gas sample. On crossing the finish line, the watch was stopped, the fuel feed switched from the measured tube to the reserve supply, and the gas-sample tube closed. The repeated acceleration was required, according to the final report, "to consume an accurately measurable quantity of gasoline and to provide a representative gas sample."[13]

The investigators' report noted, "While there was a fair amount of traffic, no trouble was experienced from stoppage by other vehicles due to the absence of much traffic on the few intersecting streets."[14] One can only imagine, however, what the drivers of other vehicles thought about a car or truck bearing a big sign saying, "Government Test Car, U.S. Bureau of Mines," with three men in it wearing white lab coats as it lurched down the street, speeding up and then abruptly stopping.

The tests were so meticulously planned and executed that there was even consideration given to the differences in gasoline obtained in Pittsburg to that available in New York. After obtaining and testing numerous samples from both places, the engineers determined that New York gasoline was likely to produce slightly lower percentages of carbon monoxide, but "practical considerations" indicated that there would not be consistent differences in exhaust-gas composition due to the quality of gasoline used.

The average results of the tests provided by the final report, delivered November 1, 1920, showed carbon monoxide production varying from two cubic feet per minute for the largest trucks to a little over one cubic foot per minute for passenger cars. This was more than what had been anticipated, and the tunnel engineers began to fear that the ventilating plant would have to deliver more fresh air than originally thought necessary. This would potentially entail much larger blower equipment and a greater annual expense for electrical power than had been estimated.

Nonetheless, this data, when combined with the projections concerning the number and types of vehicles likely to use the tunnel, afforded the engineers a baseline for determining the amount of carbon monoxide that would be produced in the tunnel when operating at full capacity. But before the engineers could design a ventilation system to reduce or remove that gas, they had to know how much carbon monoxide a motorist could inhale without an adverse affect on health.

That question was answered by physiological tests conducted at Yale University. The first test involved locking up individual student test subjects in a gas-tight chamber of 226-cubic-feet capacity (about the size of a small closet), into which varying amounts of carbon monoxide were introduced. At different intervals, the subject would shove his or her hand through a self-sealing elastic panel and blood would be drawn from a finger for analysis. Blood was also drawn before and at regular intervals after the tests. The subjects were allowed to read while in the chamber but were also asked to execute a number of tasks, such as turning on an electric fan to diffuse the gas, standing up to look out the observation window, and opening and closing flasks to take air samples in order to replicate, as much as possible, activities similar to those in which a motorist might engage. After exiting the compartment, the subjects were tested to determine the effects of inhaling the carbon monoxide.

Another test placed several test subjects in a chamber of twelve-thousand-cubic-feet capacity with a running automobile, allowing the vehicle to exhaust its foul fumes into the chamber. Although the length of time it would take a motor vehicle to travel through the tunnel was estimated at 31.4 minutes, the subjects in both the one-person and multiple-person chambers were exposed to the fumes for a full hour, during which they inhaled concentrations of two to ten parts carbon monoxide per ten thousand parts of air breathed.

Of all the physical signs and tests of poisoning employed, headaches proved to be the most definite and reliable. The students were asked when their heads began to hurt and when the headache went away. On the basis of their responses, it was determined that no one was affected appreciably by an exposure to four parts carbon monoxide in ten thousand parts of air breathed in an hour, and only a few students reported headaches from exposure to six parts per ten thousand for the same period. Contrary to what

Waddell had claimed in the paper that had raised the ire of the tunnel engineers, it was also found that the effects of carbon monoxide poisoning were not cumulative, and all the gas in the blood stream was eliminated after a few hours spent breathing fresh air.

The engineers also tested horses and dogs, presumably because they could be gassed to a point unacceptable for a student. But since the animals could not report on how bad their headaches were or when they went away, the information obtained was less valuable. Moreover, even though the tunnel had been planned for use by horse-drawn wagons, which would pass through at a much slower speed than motor vehicles, the tests were conducted with the assumption that only motor vehicles powered by internal combustion engines would pass through at their best rate of speed. The tunnel commissions' staff engineers had already begun to realize that operational problems would ensue if slow, horse-drawn wagons were allowed to mix with motor vehicles, so they convinced the commissioners that horses would have to be banned from the tunnel. The commissioners also decided that pedestrians would be discouraged or even prohibited from using the tunnel, even though the enabling legislation specifically stated that the tunnel was to be for pedestrian use.

The tests did not consider any pollutants that might be present in vehicular exhaust other than carbon monoxide, such as lead or other heavy metals. This was partly due to ignorance regarding the harmful effects of such pollutants but also due to the assumption that carbon monoxide was the most dangerous of all and that if the percentage of that gas in the tunnel atmosphere was reduced to a safe level, then the levels of other pollutants would also be sufficiently reduced.

The tests resulted in adoption of a standard to be obtained that would provide a tunnel atmosphere with a maximum percentage of carbon monoxide of four parts in ten thousand units of air. The next task for the engineers was to determine how much air would have to be forced into the tunnel to achieve the standard of ventilation and to determine the most cost-effective means of providing that air.

Two methods of ventilation were possible: the longitudinal and the distributive. In the longitudinal method, air would be supplied at one end and sucked out the other, with a low percentage of carbon monoxide present at the entrance and a high percentage at the exit. This was, essentially, the method used to ventilate train tunnels, where people are protected from the high velocity of air by being enclosed in rail cars. It was also the method that Waddell had assumed would be used. In the distributive method, air is introduced and exhausted from the tunnel at numerous points along the length of the tunnel via ducts, which allow fresh air to be supplied at all points throughout the tunnel. This method also allows the air supply at any one point in the tunnel to be controlled by increasing or decreasing flow rates in individual segments of the tunnel, and it allows velocities to be kept within comfortable limits for motorists. The distributive method being clearly superior for a vehicular tunnel, that was the method selected.

The tunnel engineers knew that the power required to move air through the length of the tunnel increases very rapidly with an increase in duct length, so they decided to have two ventilating shafts located at each end of the tunnels, one in the river and one on land, in order to break the ductwork into shorter segments. This would make a total of four ventilating shafts for each tube. Questions still remained, however, regarding the number and capacity of fans needed in each shaft to move the required volume of air. To answer this question, an additional contract for further tests would have to be signed with the Bureau of Mines.

On July 30, 1920, shortly after the tunnel engineers sent their letter about the Waddell paper to the ASCE, the Board of Consulting Engineers approved the contract, plans, and specifications of Contract No. 1. The commissions gave their approval about three weeks later, and on August 24, the contract was advertised for bids. Holland used the time available until receipt of bids to take a much-needed vacation the first two weeks of September. As soon as he returned, he reported to the commissions that contract drawings for the New Jersey land shafts were complete and that the contract and specifications for Contract No. 2 would be ready as soon as the details of the Erie agreement were settled.

Five experienced tunnel contractors submitted bids for the New York land shafts (Contract No. 1) on September 20. The bids were opened the following day. The bid of Thomas B. Bryson, vice president of Holbrook, Cabot and Rollins Corporation, was lowest, at $650,802.50, just $57,811 more than the estimate prepared by Holland's staff. The commissions awarded Contract No. 1 to this firm on October 1, with a term of one year.

Both of the New York land shafts would be located between Washington and West Streets, with one shaft in Canal Street and the other in Spring Street. The ventilation building housing equipment for both tubes would be located between the two shafts on a fifty-foot strip of land to be acquired from New York City. This property had once been the location of the Clinton Market and was now used by the municipal street-cleaning department. The Spring Street shaft would be 54 feet deep, while the Canal Street shaft would be 60 feet deep, and both shafts would measure approximately 42 feet by 48 feet. It was estimated that the work would involve excavation of 870 cubic yards of earth above mean high water, 8,330 cubic yards of earth below mean high water, and about 10 cubic yards of solid rock.

The contractor would also have to relocate and reconstruct about two hundred feet of the old Canal Street sewer, and it was known from the start that this would be a difficult and expensive job. The street was named for a forty-foot-wide canal built between 1805 and 1808 to drain what remained of Collect Pond, a small spring-fed lake located roughly where the Hall of Justice and Tombs Prison now stood. Over the years, the pond had been filled in with construction debris, trash, and dirt from a nearby hill, but nothing had been done to allow the water to drain away until construction of the canal. Beginning about 1815, the canal was also filled in and eventually covered over and converted into the largest sewer in the city. Near the shaft site, the sewer became an outlet for many lateral sewers, and it served a large part of Lower Manhattan.[15]

The first event in execution of Contract No. 1 was the groundbreaking on October 12, 1920. The ceremony was held in Canal Street Park, also known locally as Three Corners Park, a small plot of land bounded by Washington Street on the east, West Street on the west, and Canal Street, which split and wrapped around it on the north and south. Construction of a grandstand and a speaker's stand had been rushed so that the event could take place as part of the general celebration of Columbus Day.[16]

A unit of cavalry from the Essex Troop, otherwise known as the First Cavalry of the New Jersey National Guard, and a battalion of the Fourth Regiment of the New Jersey National Guard escorted New Jersey Governor Edward I. Edwards, U.S. Senator (and former governor) Walter Edge, Jersey City Mayor Frank Hague, and assorted other New Jersey officials to the Jersey City waterfront, where a special ferryboat took them to the Twenty-Third Street landing and from there to the New York National Guard Armory at Park Avenue and East Thirty-Fourth Street. At the armory, they were received by New York commission Chairman George R. Dyer, who wore a top hat and carried a bamboo walking stick for the occasion.

New York was also represented by Lieutenant Governor Harry C. Walker, standing in for Al Smith, who had to cancel his scheduled appearance at the last moment. Dyer, Walker, U.S. Senator William M. Calder, New York City Mayor John F. Hylan, and other assorted officials from New York had been escorted to the armory by mounted units of the New York

City Police Department and by the Seventy-First Regiment of the New York National Guard, commanded by former New Jersey Interstate Bridge and Tunnel Commission member Colonel J. Hollis Wells.

Following a buffet lunch at the armory, the official party boarded automobiles, with Edwards, Calder, and Walker in one and Edge, Hague, and "his retinue" in the next car. The procession left the armory about 1 p.m., accompanied by the music of police and National Guard bands. The group drove slowly west and then south from the armory to the park, waving to clusters of citizens along the route as they passed. Flags flew and bunting hung from many buildings along Canal Street, where residents leaned out of windows and stood on roof tops to observe the parade. Along the river front, hundreds of trucks, idle on the holiday, afforded bleacher seats for the children whose playground had been taken over by the project. In addition to the approximately one thousand invited guests in the grandstand, many more people had assembled in the streets and on the sidewalks.

Dyer, who presided at the ceremonies, introduced Monsignor Michael J. Lavelle, rector of St. Patrick's Cathedral, who offered prayer for the success of the undertaking and the safety of those engaged in it. Then Walker, delivering Smith's speech as his representative, declared that the vehicular tunnel was the ring which symbolized union of the interdependent states of New York and New Jersey. He also noted that while the groundbreaking was essentially an interstate event, "no one misses the idea that the tunnel, in aiding the further development of the Port of New York, will have to be classified as a national institution." Citing the long-established need for the tunnel, he added, "It is not so surprising that the tunnel is about to be constructed as it is that it was not constructed long ago."

Governor Edwards followed, characterizing the tunnel project as "the greatest in the world at the present time." He also said that the tunnel would foster the business and friendly relations of the two great states it connected, and he praised the two state commissions for having brought the project to a realization. Senator Calder was the next speaker, saying that no other public work undertaken in that part of the county had contributed more to the future of New York and its environs and that it had become a real necessity to the business of cities and towns in both states. Senator Edge then had his turn, telling the crowd of his interest in the project as governor and in his present position as senator. He said the project was a concrete demonstration of how state governments could be helpful to each other, and then he spoke of his surprise that "the antiquated methods of a century ago are still in use. The day of the ferry is fast coming to an end and soon we will have more tunnels of this kind." Mayors Hylan and Hague also spoke, pointing out the advantages the tunnel offered to their respective communities. They were followed by Wilgus, who presented the engineers' perspective.

Speechmaking done, the party, led by Edwards and Walker, made their way to the enclosure where digging the first shovelful of earth would mark the official commencement of construction. Edwards grabbed a silver shovel and Walker a silver pick, and they stood with a group of other officials as photographers took pictures. Then one of the newsreel photographers said, "Let's have some action." Edwards handed his hat to an associate, Walker spat on his hands, and the two men began moving dirt from ground to wheelbarrow with great vigor while the newsreel photographers hand-cranked several yards of film through their cameras. Whenever Edwards and Walker slowed, photographers asked for more action. The wheelbarrow was nearly full when, in the background, the Seventy-First Regiment struck up the "Star-Spangled Banner." The two men then dropped their tools and breathed hard as they and everyone else stood at semiattention, their bare heads unprotected against a chill wind from the Hudson River. The anthem completed, the two officials resumed digging, until a jokester asked, "Hey, have you two got your union cards?" At that, the digging for the day ended.

It took the contractor almost another two weeks, until October 27, to begin excavation of the shaft sites. Six days later, New Jersey voters approved the bond issue for financing construction of the Delaware River bridge and Hudson River tunnel, and a month later, the New Jersey commission authorized an issue of $5 million thirty-year bonds at 6 percent.

With the New Jersey election past and the bond issue approved, Holland moved to strike a killing blow to the Goethals-O'Rourke scheme. At the Board of Consulting Engineers meeting on November 16, he reported on his recent trip to inspect the Six-Mile Road, Section 1, concrete-block sewer tunnel being constructed for the Detroit Department of Public Works by the Hoag and Dell Company and the O'Rourke Engineering & Construction Company. Some of the Board members had urged the trip because they thought that the sewer tunnel might demonstrate the soundness of O'Rourke's design. Holland found that the contract was in default and work had been stopped. Many of the blocks had cracked or been crushed under jack pressures, and the irregularity of the joints was such that water-tightness could not be obtained. This confirmed his opinion that O'Rourke's patented concrete-block system would not work under the Hudson River.

Holland followed up his attack at the last Board meeting of the year, on New Year's Eve, by reading a notice from the December 22, 1920, issue of the *Detroit Legal News*, announcing default of the Six-Mile Road contract and of O'Rourke's contract for construction of the Joy Road Public Sewer. It was the Board's consensus of opinion that the matter should be brought to the attention of the commissions.

Holland also put another nagging issue to rest in December. In the first week of the month, he told the Board that it was necessary to make a decision in regard to adopting the trench method as an alternative to the shield method, so that contract plans could be prepared for advertising the river portion of the work. He had concluded that the trench method would result in serious risks of accident to the floating facility and to the tunnel sections, with resultant casualties, claims for damages, and delay in the completion of the project, unaccompanied by savings in either estimated costs or time of construction. His findings were accepted, and he was allowed to drop any further consideration of this method.

Following review of the official ventilation-study reports of the Bureau of Mines, which were delivered in November 1920, the Board and staff engineers also decided to execute a new contract with the Bureau for tests to determine the best method for moving the required quantity of air through the tunnel. The agreement called for construction of concrete ductwork, similar to that to be used in the tunnel, at the Bureau's Central District Experiment Station at the University of Illinois in Urbana. Special three-hundred-horsepower fans would be installed to move air through the ductwork. The cost of these experiments would be shared by the Bureau because of the fundamental importance of the data obtained to the mining industry, where air was often forced through tunnels at high velocities.

Another part of the agreement called for construction of a model tunnel at the Bureau's Experimental Mine at Bruceton, Pennsylvania, just outside Pittsburgh. A four-hundred-foot-long oval-shaped tunnel having ducts similar to those proposed for the vehicular tunnel would be constructed underground, and vehicles would be operated in it so that data could be obtained on diffusion of gases and effects on motorists.[17]

The engineering staff completed plans and specifications for Contract No. 2 in December 1920. This contract would cover construction of the New Jersey land shafts in the Erie Railroad yards and about three hundred feet of approach construction west of each shaft. The anticipated date for contract award was set at March 15, and the date of tunnel completion was now projected to occur sometime in early summer of 1924.

On February 28, 1921, five days after the commissioners authorized printing the contract-bid documents, Governor Edwards appointed Frank F. Gallagher of Collingswood, New Jersey, Charles S. Stevens of Cedarville, New Jersey, and Adams (as a reappointment) to the New Jersey commission. Stevens replaced Daniel F. Hendrickson, whose term had expired, and Gallagher replaced Richard T. Collings, who had died October 4, 1920. The commissioners soon elected Adams chairman and Shay vice chairman at the annual reorganization meeting.

On March 8, 1921, the day after the commissions advertised Contract No. 2 for bids, they held a joint meeting to discuss groundbreaking ceremonies to be held on April 16 in Jersey City. Just before adjournment, Adams and Boyle informed their associates from across the river, "in a most casual manner," according to Dyer, that the New Jersey negotiating committee had committed both commissions to pay for the widening of streets in Jersey City. The cost would be about $1 million, with New York paying half. "We were amazed," Dyer later claimed, "and so told the New Jersey Commission. We further told them that this was the first time that we had ever heard of such a promise."[18]

The New Jersey commissioners insisted, however, that the specifics of the agreement were widely publicized in Jersey City shortly after the deal was made. In addition, the New York commissioners had authorized the New Jersey negotiating committee to act for them and were thus bound by its actions. The fight that ensued between the two state commissions over whether the tunnel project was obliged to pay for street widening erupted into a scandal involving destroyed meeting minutes, secret real estate deals, intentional misstatements to the public, and dismissal of the Board of Consulting Engineers. Whatever merit there may have been in the argument that the project should pay the cost of improvements beyond the tunnel portals, the potential benefits of an objective study of the problem were negated by political considerations and subjective self-interest. This shortsightedness was compounded by lack of experience, on the part of engineers and politicians, regarding the manner in which road improvements across state boundaries should tie into regional highway systems.

Above, left: Walter E. Edge; *above, right:*
George W. Goethals; *left:* John F. O'Rourke

New York and New Jersey commissioners executing agreement to undertake the construction of the tunnel, 1919. *Seated, left to right:* Barlow, Adams, Collings, Boett-ger, Campbell, Richards, Noyes, Dyer, Bloomingdale, Williams, Whalen, Shamberg, Hawkes, Windels. *Standing, left to right:* Barradale, Holland, Frohlich.

Left: Clifford M. Holland; *right:* Ole Singstad

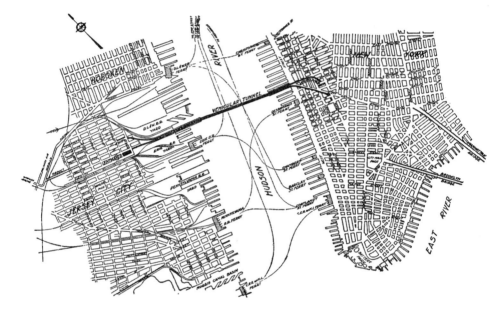

Location map

Top: site of the tunnel, looking west from New York City; *bottom:* site of the tunnel, looking east from Jersey City

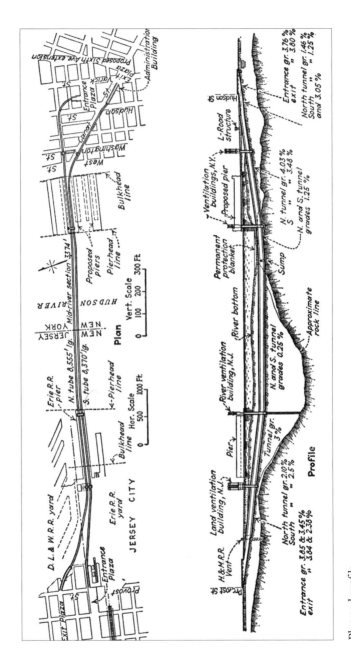

Plan and profile

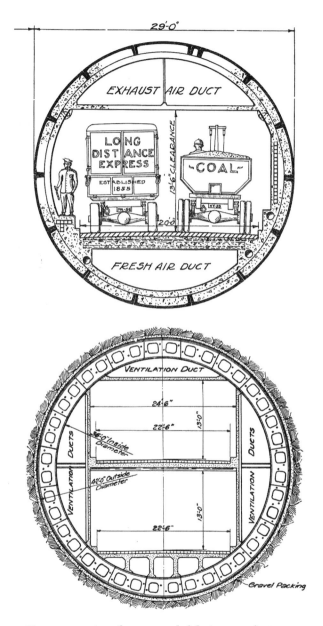

Top: cross-section of recommended design, 1919; *bottom:* cross-section of Goethals-O'Rourke design

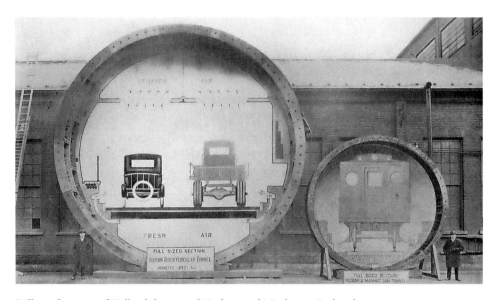

Full-sized section of Holland design and Hudson and Manhattan Railroad section

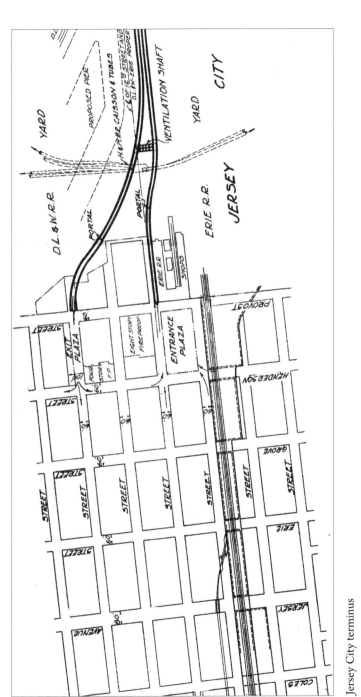

Jersey City terminus

Top, left: assembling shield in Canal Street shaft; *top, right:* land-shaft caisson at Spring Street; *bottom:* shield, south tunnel, Canal Street at West Street, New York City

Launching of New Jersey north river caisson, Staten Island

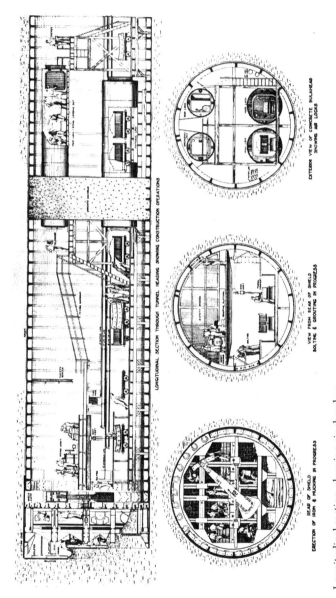

Longitudinal section, showing air chamber

Man lock, south tunnel, New York

Right: erector arm;
below: sandhogs waiting
to go into Canal Street
shaft, south tunnel,
New Jersey

Top: hauling a car of muck out of a muck lock, south tunnel, New Jersey; *bottom:* muck coming through doors in shield, south tunnel, New York

Top, left: men tightening bolts; *top, right:* grout nose and nipple in plate; *bottom:* holing through

7

Another Long and Costly Delay

Recent developments indicate that another long and costly delay threatens the construction of the Manhattan–Jersey City vehicular tunnel under the Hudson River as far as work on the Jersey side is concerned.
—*New York Evening Post*, September 22, 1921

ON MARCH 8, 1921, the same day that Adams and Boyle informed the New York commissioners of the commitment regarding street widening, George Watson, the only member of the Board of Consulting Engineers originally from New Jersey, read a letter to the Board stating, "To my mind the widening of the streets to give ample plaza and exit facilities on each side of the river is essentially a tunnel project." Watson then noted that the city streets in New York were already wide enough to handle tunnel traffic (or so it was thought). But if they had not been, he claimed, the tunnel project would have been responsible for their improvement, just as it was responsible for improving Jersey City streets.[1] After reading his letter, Watson moved that the Board pass a resolution advising the commissions that Holland's street plans for the Jersey City approach were inadequate and that the commissions should consider widening the affected streets. Burr, Davies, and Wilgus voted no, and Bensel abstained. It was the consensus of opinion that Watson should take this question up with the commissions, as the majority of the Board did not think that this was an engineering matter.

Watson did not let the matter rest there. On March 14, he sent a letter to both commissions stating that they should "study and consider the problem of traffic congestion that will necessarily follow on the New Jersey side of the river, and that its effect upon the tunnel due to the inadequate width of the streets of the one side of the river when compared with those of the other; and that they recommend a plan that will give equal entrance and exit facilities for the necessary traffic on either side."[2]

Watson's letter was not well received by the New York commissioners. Although they agreed that street improvements would be needed in Jersey City to accommodate tunnel traffic, those improvements were the responsibility of the State of New Jersey, Hudson County, or the City of Jersey City. The tunnel project terminated at the entrance and exit plazas, and any improvements beyond those points, whether necessary as a result of tunnel construction, general population growth, urban development, or any other reason, were beyond their purview.

Hague, for his part, let it be known that if the New York commissioners did not abide by the street-widening agreement, Jersey City would not vacate Eleventh Street and the Erie Agreement would collapse. This, in turn, would make it impossible for Contract No. 2 to be let, and all work on the Jersey side of the river would be held up. The New York commissioners countered by threatening to take the land through condemnation proceedings. They apparently assumed that they could count on the support of the New Jersey commissioners from southern counties, which had no political stake in the street-widening issue and would want to force a resolution of the problem before it imperiled their plans for the Delaware River bridge. This counterthreat was somewhat of a bluff, however, because the New York commission had no power to condemn land in the state of New Jersey, and it was unclear if the New Jersey commission had that right either, or would exercise it if they had it. Moreover, condemnation proceedings would take a long time and have to go through the New Jersey courts, where success was far from assured.[3]

The street-widening controversy was more than just an argument over what negotiating authority the New York commissioners had given to their brethren from across the river or if they had known all along about the commitments made. It also became a debate among the New Jersey commissioners over the true purpose of the agreement with Hague. At a later joint commission meeting on April 26, 1922, more than a year after the issue first surfaced, Jersey City Corporation Counsel Thomas J. Brogan, a Hague loyalist, reminded the commissions of the promise made by the New Jersey negotiating committee in August 1920 to widen the streets. New Jersey Deputy Attorney General Emerson Richards, who had been counsel to the New Jersey commission at the time and had been present at the relevant meetings, angrily said to Brogan, "You know that that promise was only for the consumption of the public. It was not the real thing." To which Brogan replied, "You mean to say that the Tunnel Board [meaning, the New Jersey commission], when it made a pledge, was just handing out pure bunk? Let us know then, how much of what the Tunnel Board does is bunk and what is on the level. I think that the public has a right to know just to what extent it is being buncoed by false declarations by the Tunnel Board." Having no good response, Richards dropped the subject.[4]

Apparently, at least some of the New Jersey commissioners believed that the deal made with Hague and the city commissioners was a sham, never to be honored and created simply to allow them to present themselves to voters as champions of street improvements. Presumably, if the minutes of commission meetings were properly kept, they would shed some light on the true intent of the agreement and reveal who said what and when. Some of those minutes, however, had been destroyed by E. Morgan Barradale, secretary of the New Jersey commission. After being attacked by Adams over destruction of those documents, Barradale resigned at a commission meeting on March 22, 1921. Other commissioners asked him to withdraw his resignation, and one of them called for a resolution of support for the beleaguered secretary. The motion failed, the resignation was accepted, and Barradale was replaced by James P. Dolan of Jersey City.[5]

At first, no one seemed to be fully aware of just how long the street-widening controversy would hold up the award of Contract No. 2. Holland was more concerned with a rumor that the contract would be awarded immediately upon opening of bids, without review by the engineering staff or the Board. In other words, some people had the impression that the fix was in and that the contract would not be awarded on an objective basis. Holland called a special meeting of the Board on March 22 to seek its advice, which resulted in a letter from Holland to the commissions. The minutes of the Board meeting do not reveal what his letter said, but it may be safely assumed that Holland would accept nothing less than award of the contract on the basis of objective review procedures.

On April 11, Holland recommended to the Board of Consulting Engineers that Contract No. 2 be let as soon as possible in order to expedite the completion of the contract as a whole. The Board so moved, with a view to having bids opened on or about June 15. With that action taken, the Board took up a variety of relatively minor construction and design issues before leaving on April 28 for an inspection trip to Urbana, Illinois, to see how the new ventilation experiments were going. Bensel, Burr, Davies, and Watson returned to New York on May 2, having left Holland in Urbana. The tests were going well, and the Board soon recommended additional tests of full-size models. But the street-widening controversy was about to blow up in their faces.

On May 3, the New Jersey commission passed a resolution stating that whereas an "unfortunate misunderstanding" had arisen between the commissions in regard to street widening in Jersey City, and this misunderstanding "may justly be attributed to a failure upon the engineering staff and Consulting Engineers to give full and proper consideration to the plan as a whole, and to the traffic conditions in the vicinity of the Jersey City entrance and exit," the New Jersey commissioners had decided to dismiss the Board, effective May 14, and appoint a new three-member board to advise them on all technical matters. This board would meet and confer with a similar board to be appointed by the New York commission.

On the same day the New Jersey commission adopted the dismissal resolution, the New York commissioners received more unsettling news. When the street-widening controversy flared up in March, they were led to believe that many New Jersey political leaders would support their stand that widening of streets in Jersey City should be seen as a local improvement and thus not subject to being financed by the tunnel project. But they found out that the political winds were now blowing in a different direction, and many state leaders in New Jersey were anxious to have city streets widened for a full six blocks west of the tunnel entrance to conform to the Lincoln Highway, then being completed across the state.[6]

On December 13, 1913, only three months after the Lincoln Highway Association first announced the official route of the highway, the newly reconstructed section between Jersey City and Newark formally known as the Plank Road, because of the heavy wood timbers that made up its surface, was dedicated as the "Essex and Hudson Lincoln Highway." Nearly one hundred feet in width, with two twenty-foot roadways, one paved with wood blocks and the other with granite, the road was the first newly constructed and christened portion of the Lincoln Highway in the country.[7] But in 1921, the Lincoln Highway through Jersey City still ran through narrow city streets built to a local standard inadequate for the ever-increasing long-distance traffic.

The Jersey City Board of Engineers and some state politicians had decided that once the tunnel was complete, the Lincoln Highway in New Jersey would begin in the middle of the Hudson River. Designation of the Jersey City streets leading to the tunnel and the tunnel itself as parts of an interstate coast-to-coast highway strengthened the position of Hague and his acolytes that the necessary street widening was not a local improvement, to be paid for with local funds.

In a full-page advertisement placed by campaign chairman John Boyle in the May 4, 1921, edition of the *Jersey Journal*, Hague and the other Jersey City commissioners seeking reelection promised "to make Henderson Street, at present a narrow and congested thoroughfare, into a 100-foot street from the Hoboken line to the Morris Canal; and also to make a wide thoroughfare leading from the entrance of the tunnel to the hill section, and to widen several of the streets for several blocks adjoining the entrance to the tunnel. Upon the proper treatment of the Jersey City end of this great tunnel and its approaches will depend in a large measure the advantages to be reaped by Jersey City from this improvement."[8] The ad did not state how the "Hague slate" of candidates intended to pay for the improvements, but they passed the word around town that they would squeeze the money out of the tunnel project.

As far as the New York commissioners were concerned, since the U.S. Senate Committee on Interstate Commerce voted down their request for federal funding in December 1918, there could be no more consideration that the tunnel was part of a national highway. They maintained that the tunnel stopped at the entrance and exit plazas in each city, and the enabling legislation did not permit them to spend funds for improvements beyond the plazas. Hague, the Jersey City commissioners, and the New Jersey tunnel commissioners argued that the tunnel project, when seen in its entirety, included portions of Jersey City streets extending several blocks beyond the plazas. They had attempted, informally, to get the Board of Consulting Engineers to agree with them. When all the members of the Board except Watson refused, the New Jersey commissioners decided to fire them all.

The dismissal was more than just an attempt to get city street improvements paid for by the tunnel project. It was also an attempt to open up new positions that would be filled on the basis of political patronage. To achieve that end, the New Jersey commission appointed St. John Clarke and George M. Wells, both from New Jersey, to a new board of "technical advisors." Clarke was an advocate for a single-tube, six-lane tunnel with a steel shell and an inner lining of concrete, to be constructed by the trench method. Wells was a partner of Goethals in the latter's consulting business, and he favored O'Rourke's concrete-block design. The commission also appointed Watson as a technical advisor, over the protests of several members who had someone else in mind. The majority of commissioners apparently saw Watson's appointment as a way to offset the patronage charge made by New Yorkers, since Watson ran his consulting business out of New York.

The plan did not work. As the New York commission and the Board of Consulting Engineers correctly informed the New Jersey commission, the compact entered into by the two states for construction of the tunnel stipulated that neither commission could act independently. Since the New York commission refused to go along with the dismissal, the action of the New Jersey commission had no legal force or effect. The Board continued to go about its work, even though the New Jersey commission refused to accept its reports or to approve payment of salaries to its members.

Another challenge to the New Jersey commission's action arose on May 17 when Watson sent a letter refusing to accept his appointment. To his credit, he stated that he still considered himself a member of the Board, as originally and jointly appointed. The New Jersey commissioners reacted by releasing a statement claiming they were glad Watson refused to accept because he clearly was not in sympathy with their position. Henceforth, they would refuse to have anything to do with any members of the Board of Consulting Engineers.[9]

While the commissioners from both states dug in their heels and refused to yield their positions, rumors began to circulate about their real intentions. Some people speculated that there was a secret plan to get both commissions dismissed by the legislatures in each state so that the newly created Port of New York Authority could take over the project. The bi-state compact authorizing creation of the Port Authority had been signed on April 30, 1921, and it was widely expected that the compact would soon receive the consent of Congress and be signed by President Harding before the end of August. Others, including the New York commission's Chairman Dyer, thought that Hague, who was rumored to have financial interests in concrete sales, wanted to force a reconsideration of the Goethals-O'Rourke design. Boyle defended his good friend against that charge by stating, "Mayor Hague has no more to do with this concrete scheme than has the man in the moon."[10] But a new explanation for the holdup soon came to the public's attention, thanks to a series of investigative reports by the *New York Evening Post*.

On May 25, 1921, the *Post* published the first of several articles uncovering the self-serving real estate dealings of Adams, his relatives, and his business associates regarding land near the mouth of the tunnel. "As the situation stands now," the article stated, "the chairman of the New Jersey Commission is leading the fight in the present dispute, which is holding back construction of the tunnel, despite the fact that the question at issue between the two bodies centers around property in which the New Jersey chairman is interested, and any decision in that case will affect, at least indirectly, the business of the cold storage company of which he is the head and in which he has a large financial interest."[11]

The article then cited, in great detail, Adams's pivotal role in securing the tunnel's enabling legislation in both states and the myriad real estate dealings in which he and his associates had engaged after he had been appointed to the New Jersey commission. When questioned by the *Post* reporter, Adams downplayed his interest in the various companies affected and claimed that the tunnel would cause land near its mouth to depreciate rather than increase in value. There would have been no point, therefore, for him to buy real estate in the vicinity. But in a previous statement setting forth the New Jersey commission's position in the street-widening dispute, Adams had stated, "The proper provision for the necessary entrance and exit plazas should be made before the advance in property values, which will occur after the opening of the tunnel." When reminded of this statement, Adams dismissed the whole question of his real estate dealings by saying that if he considered land near the tunnel's mouth a good buy, he would buy it tomorrow and see nothing unethical in doing so simply because he was chairman of the New Jersey commission. "There would be nothing inappropriate in my doing this," he claimed, "and any business man will tell you the same thing."[12]

In an article the following day, the *Post* disclosed Boyle's ownership of stock in the Union Terminal Cold Storage Company. The paper also noted that Boyle was campaign manager of the "Hague machine" during the election earlier in the month when the Hague slate of candidates had promised to widen the streets in question.[13]

The reaction of other newspapers was predictable. Those based in New York reported the story as an unfortunate development, while also expressing the hope that New Jersey interests would clear the matter up before it did any more harm to the project. The New Jersey–based papers took a stronger stance, claiming that hidden foes of the tunnel project, cowards and slanderers all, were behind the "false" charges.

Governor Edwards reflected the opinion of most New Jersey Democratic leaders in stating, "The story is propaganda from an interested source. It can probably be traced, and if it is traced appropriate action will be taken." He also referred to Adams's denial of wrongdoing, saying that the facts were as Adams had presented. "He is doing what any red-blooded man would do and ought to do in his position," Edwards maintained.[14]

With rumors continuing to circulate that the tunnel project might be

taken away from the state commissions and given to the Port Authority, great pressure was applied to the commissioners to settle the street-
widening controversy. After a joint conference on June 7, it seemed that
a compromise had been reached. The new plan called for Twelfth Street
to be widened to serve as an entrance plaza, with the tunnel project paying the cost of widening from Provost Street to Grove Street. From Grove
Street to Jersey Avenue, Twelfth Street would be widened at the expense of
the Erie Railroad. The tunnel project would also pay the cost of widening
Fourteenth Street from Provost Street to Henderson Street so that it could
accommodate the exit plaza and would pay for widening Henderson Street
between Twelfth and Fourteenth Streets. Jersey City would pave a portion
of Twelfth Street from Grove Street to Jersey Avenue, but that would be its
only expense.[15] All that remained to settle the impasse was agreement by
the Erie Railroad, which the commissioners immediately sought.

The New York commissioners, anxious that nothing stop the planned
advertising for bids on Contract No. 2 and fearful that the project might
be given over to the Port Authority, had apparently decided to make a
major concession. Although Hague did not get all that he wanted, he received more than the New York commissioners had initially seemed willing to give. This led the *Jersey Journal* to pronounce that the charges made
against the New Jersey commissioners had been only "propaganda that
usually creeps into fights between public bodies—showy and all that but
harmless." The *Journal* also knew to whom the credit should go for bringing
forth an agreement. "All that's now left to do is to build the tunnel," the paper crowed, "and be thankful that on the Tunnel Commission there is such
a level-headed and clear-sighted Commissioner as John F. Boyle of Jersey City."[16] Whatever the motivations of the New York commissioners, it
seemed that their capacity for compromise would lead soon to an opening
of bids for Contract No. 2 in July. The project could again move forward.

One problem still to be worked out, however, was the illegal dismissal of
the Board of Consulting Engineers. At the end of June, Wilgus announced
his intention to sue the New Jersey commission for unpaid wages. His position was strengthened by a report released in the middle of July by the
New York chapter of the American Association of Engineers, condemning
the dismissal as unethical and illegal and finding that it tended to hurt the
entire engineering profession.[17] The New Jersey commissioners held fast
to their decision, however, and refused to accept reports from the Board.
They focused their energies instead on plans for a great groundbreaking
celebration to be held in Jersey City on Sunday, August 6. Hague appointed

a general citizens' committee to organize the event, which would consist of parades, numerous bands, speeches, and, it was hoped, an appearance by President Harding. Nearby towns, including Bayonne, Paterson, Newark, Passaic, and Elizabeth, announced their intention to participate in the festivities, which were expected to be grand enough to attract national attention.[18]

The groundbreaking had to be delayed, however, due to difficulties with the award of Contract No. 2. Bids were received on July 12, 1921, with Holbrook, Cabot and Rollins (the winning bidder for Contract No. 1) the lowest bidder. Presumably, this firm could bid the job fairly low due to the economy involved in already having staff and equipment at work on the New York side of the river. The excavation of the Spring Street shaft would be complete by July 15, and excavation of the Canal Street shaft was expected to be completed early in August. But the firm had not accompanied its bid with the required surety check, which was delivered twenty minutes after expiration of the time limit. The messenger had simply forgotten to bring the check with him, and by the time it arrived, it was too late. To avert a threatened lawsuit by the second-lowest bidder, the Jocylen Construction Company of Bayonne, New Jersey, all bids were rejected, and a new date for bid opening was set for the following week.[19]

The Shaft Construction Company of New York was the lowest of six bidders when the new bids for Contract No. 2 were opened on Wednesday, July 19. Its bid was $369,815.50, about $5,000 under the next-lowest bidder, Holbrook, Cabot and Rollins. The Shaft Construction Company's bid was also $173,239.50 less than Holland's estimate of the cost to do the work. Holland requested that Thomas E. Shea, the president of the company, come by his office the next day for a conference regarding the company's qualifications and plans for carrying out the work. When Shea arrived, Holland found it impossible to discuss the contract with him because Shea was an attorney and not an engineer or contractor, and Shea stated that he was responsible only for the financial arrangements. Holland asked him to come back the following day to meet with Windels and Freeman and to bring with him someone who could discuss the plan of construction.

On Friday, Shea returned with John O. Devlin, the commissioner of public safety in Bayonne, New Jersey, who presented himself as vice president of the company. Along with Shea and Devlin, the shareholders of the company were Devlin's brother, William O. Devlin, a contractor from New York; John B. Cusick, a merchant from Bayonne; and James T. Brady, president of James Brady Supply Company. Also in attendance was the engineer whom Devlin stated he would retain to direct the work, John F. O'Rourke. The O'Rourke Engineering & Construction Company had submitted a high bid on July 12 but had not bid on July 19. Apparently, O'Rourke decided at some point after July 12 that his best chance for profiting from the project was by forming an alliance with contractors who were politically connected enough to win the construction contract. According to Devlin, O'Rourke had no ownership interest in the Shaft Construction Company and was merely a subcontractor.

The conference on Friday, together with a conference the following morning, revealed that the Shaft Construction Company had been incorporated on the day the bids were received, and none of the officers or stockholders had any experience in the type of work that it proposed to do. The capital of the firm, as stated in the Articles of Incorporation, was a mere $10,000, and its statement of assets showed a total of $113,500, consisting of $78,500 in cash and checks and $35,000 worth of equipment.[20]

If Holland thought the bid tainted, he had good reason. Not only were the officers and incorporators of the company inexperienced, it was likely that they planned to funnel kickbacks to those, such as Adams and Boyle, who might be in a position to help them win the work. O'Rourke certainly was not beyond making such payments. In the latter part of 1920, he gave a $5,000 and a $3,500 check to a man who later endorsed the $5,000 check to another man, who made payments to the campaign fund of Queens Borough President Maurice E. Connolly. At the time, O'Rourke sought a contract to build concrete-block sewer tunnels for the borough. When O'Rourke was asked during an investigation by a committee of the New York State legislature on October 6, 1921, for "his ideas" on what he got for his money, he replied, "My ideas are that anybody who can get me a contract in ten seconds, I will pay him $8,500 for ten seconds."[21]

Holland, as might be expected, recommended to the Board at its meeting on July 22 that the bid of the Shaft Construction Company be rejected because of their lack of experience and that the contract be awarded to the next-lowest bidder. The Board agreed, pending a ruling from Windels that the bid could be voided. After Adams found out that the Board had recommended giving the work to Holbrook, Cabot and Rollins, he announced that the New Jersey commission would insist on award of the contract to the Shaft Construction Company.

Whether Adams had any personal interest in the Shaft Construction Company is unknown. But the project was also held up by delays in the signing of the Erie Railroad agreement, and these delays were the direct result of a new attempt by Adams to benefit, personally, from the project. The Erie Railroad ran a small section of tracks, a "siding," to Adams's Union Cold Storage Terminal Company building, and that siding stood on land needed for the tunnel's Jersey City entrance plaza. Although Adams had a license from Jersey City for the siding, he had no contract with the railroad guaranteeing a continuing right to the siding. He therefore had James P. Dolan, secretary to the New Jersey commission, place clauses into a draft of the Erie Railroad agreement recognizing that his company had such a right. Adams hoped that these clauses would give him the right to claim payment for heavy damages suffered by loss of the siding. When Windels found out about the insertion, he demanded that the clause be taken out.

The commissions and the railroad agreed that a proof of the agreement, without the siding-related clauses included, should be sent to the printer for typographical corrections. There were two official copies of this proof: one held in the office of Erie Railroad Chief Engineer Robert C. Falconer and one held by Holland. One week before the proof was to be sent to the printer, Adams demanded that Holland turn over his copy, saying that he would see that it got to the printer. Holland refused. Adams then obtained a duplicate, inserted the clauses, and sent the altered copy to the printer. Adams also demanded that the clauses be inserted into the official proof held by Holland and threatened to have the entire engineering staff fired if they did not comply. One again, his demand was denied. The New York commissioners also contacted the printer and had the clauses removed, without telling any of the New Jersey commissioners.

When the New York commissioners delivered the official proof contract to the New Jersey commission on September 15, 1921, and Adams saw that his clauses had been removed, he took the bound copy of the contract to Peter J. Carey Print, a printer of which he was an officer and stockholder. Certain pages were removed, new pages inserted with his clauses added, and the rebound contract presented to the New York commissioners "with impressive formality."[22] When the New York commissioners read the document and saw what Adams had done, they once again demanded that the offending clauses be excised. Boyle, in response, told them that if the clauses were not part of the agreement, there would be no tunnel. Once again, there was a stalemate.

At least the ventilation studies were going well. By July, the elliptical test tunnel was under construction in Bruceton, Pennsylvania, about 135 feet below ground in an abandoned coal mine. The tunnel was formed by uniting two existing parallel shafts of approximately 400 feet in length with curved concrete sections. Inside the tunnel, eight automobiles would be driven around and around inside a rectangular box, nine feet wide and seven and a half feet high. Underneath the floor there was an air duct four feet high, and above the ceiling there was another duct five feet high. Either duct could be used for providing fresh air or exhausting polluted air. Sampling tubes were located at eight stations, set five feet apart along the length of one straight leg of the track. These tubes connected to a central station from which air from all of them could be pumped. Each vehicle used in the experiments would also be loaded with testing and sampling equipment.[23]

With the test tunnel under construction, Holland made plans to make an inspection trip, beginning August 13, to study vehicular tunnels in England, Scotland, France, and Germany. Because of limitations in length and use, none of these tunnels was similar to what he intended to build under the Hudson River. But he wanted to learn anything he could that would inform the ventilation design or help him select the best lining material for the interior of the tubes. He returned to New York on September 26, convinced that Singstad's preliminary ventilation design was on the right track. The tests in Pennsylvania would provide the final confirmation. With that hope in mind, Holland, Burr, Davies, Williams, Noyes, Shay, Gallagher, Stevens, and New Jersey Legal Counsel Richards made an inspection trip to Bruceton, Pennsylvania, on Thursday, October 20. Apparently, the New Jersey commissioners who took the trip were still willing to work with the "dismissed" Board members.

All of the men spent two hours in the underground test tunnel, undergoing the same routine as other test subjects. The medical staff took each man's temperature and pulse rate before and after he went into the tunnel, with blood also taken upon exiting to measure carbon monoxide levels. Entrance into the test track was made through air locks, which assured that the air inside would not be disturbed by people coming and going. Inside the tunnel, the men rode around in the test vehicles for forty minutes. At the conclusion of the vehicular tests, the visitors watched a demonstration of how quickly the tunnel atmosphere could be exchanged. Test personnel set off smoke bombs and then pumped the smoke out of the test tunnel. Optical instruments allowed the ventilation engineers to measure the density of smoke by measuring the percentage of absorption of green light through fifty feet of tunnel atmosphere.[24]

The engineers and commissioners returned from their trip convinced, finally, that the tunnel could be safely and economically ventilated. To help convince others, the engineers installed a model ventilation system in their office at the Hall of Records in New York. The three chambers of the model —one representing the traffic tube, one the fresh air duct, and one the exhaust duct—were enclosed in a framework with glass ends. For the edification of visitors, smoke from bombs would be driven into the lower chamber and sucked out through the upper chamber, with the middle chamber quickly cleared.[25]

Resolution of the ventilation issue allowed Holland, his staff, and the Board of Consulting Engineers to shift their focus to the work being done under Contract No. 1 and to the innumerable smaller but important challenges still to be met. As they waited for the commissioners to work out their dispute and settle the Erie Railroad agreement, they made decisions about the type of material to be used in paving the road surface (granite), the type of material to be used in finishing the interior walls (tile), the type of lighting to be installed, the warning and directional signage to be used, and many other details.

On November 15, the New Jersey commissioners, apparently realizing that they could not force award of Contract No. 2 to the unqualified Shaft Construction Company, voted to abide by the decision of the New York commission and cancel all bids for Contract No. 2. Henceforth, the work of sinking shafts in Jersey City would be combined with construction of the New Jersey land and river section tunnels in Contract No. 4. The work of driving the New York river shafts would be covered by Contract No. 3. With this constraint eliminated, details of the Erie agreement were finalized over the next few weeks, and the New Jersey commissioners affixed their signatures to the document on December 7. The New York commissioners were anxious to sign as well, but Dyer let it be known that they would not do so until it had been carefully studied, "so as to assure us that it contains nothing that might lead to future embarrassment."[26] Dyer had learned from the incident involving Adams's insertion of clauses into the Erie Railroad agreement. Following a week of review, the New York commission also approved the contract.

With the Erie Railroad agreement signed, the New Jersey commissioners resolved to erect a twenty-foot granite shaft at the Jersey City tunnel terminus in their own honor. The monument would have a base and pedestal of light-gray granite and a shaft of bronze, topped by frosted globes. On December 14, they marched down the hall to the offices of the New York commissioners in the Hall of Records, presented their plans, and asked for swift approval of the estimated $10,000 expenditure. They wanted the monument to be ready for unveiling at the groundbreaking ceremony in

Jersey City, now scheduled for about March 4, 1922. The New York commissioners were stunned by this display of hubris, and one of them suggested that perhaps the name and face of the commissioner who had come up with the idea be done in brass instead of bronze. They refused to pay half the cost of the shaft and ridiculed the New Jersey commissioners for having suggested it. Adams, in an attempt to save face, told newspaper reporters that the idea for the monument had originated with the New York commission, a ludicrous absurdity immediately denied by the New Yorkers. Adams vowed to push the issue, but it quietly died away.[27]

The haste of the New Jersey commissioners to have their contributions to the project memorialized may have been due to fears that they would soon be ousted. At a meeting held in Newark on December 7, the chambers of commerce of Jersey City, Hoboken, Newark, Bayonne, and Elizabeth, New Jersey, formed a committee to ask the governors of New Jersey and New York to institute a joint investigation into the work of the state tunnel commissions. The meeting was a response to a report produced by the Jersey City Chamber of Commerce that alleged the New York commission had attempted to blame the New Jersey commission for the numerous delays, with a hidden purpose of harming the interests of Jersey City. The investigation had the potential to cut both ways, however. E. W. Wellmuth, secretary of the Newark Chamber of Commerce, said that if either of the two commissions had acted in a way prejudicial to the interests of the tunnel project, the members of it should be replaced. "There is a nigger in the woodpile," he added, "but there is no use standing back of the New Jersey Commission if it hasn't any backbone."[28]

The plan for an investigation quickly began to backfire. The New Jersey commissioners found out that there were some legislators in their state who were ready to get rid of them and to appoint new commission members at the next legislative session. It was in their interest, therefore, to put their differences with the New Yorkers behind them so that the project could move forward. If progress could be made, they might be able to hold on to their appointments. With this in mind, Adams and Boyle were persuaded to give up their demands that a clause be included in the Erie Railroad agreement regarding a siding to the Union Terminal Cold Storage Company.

With all plans, contracts, and specifications for Contract No. 3 and Contract No. 4 ready by the end of the year, bid documents advertised on December 29, and the Erie Railroad agreement signed by both commissions, it looked like bids could be opened by February 7, 1922, and groundbreaking in New Jersey could take place early in March. This would allow for completion of the tunnel by December 31, 1924. Two hundred sets of bid documents were sold to potential bidders, but it was anticipated that no more than five bids would be received because of the limited number of firms that could take on such massive contracts. According to the *New York Times*, the contracts, to be let jointly, would probably be the largest ever for a public work project in the United States and would employ approximately seventeen hundred people. The lowest bid was expected to be in the neighborhood of $20 million.[29] In order to illustrate the magnitude of the job, Holland announced that if the washers to be used in the project were placed on top of one another, they would form an iron pillar six and a half miles high.[30]

When the date for receiving the bids arrived, only two firms actually submitted: Booth and Flinn, Ltd., and Patrick McGovern, Inc. Two other firms, Holbrook, Cabot and Rollins Corporation and Keystone State Construction Company, which intended to submit as a joint venture, said that they could not bid because of problems with obtaining the necessary surety, which had to be 20 percent of the estimated contract value. The commissions responded by rescheduling the bid opening to February 15. These four firms submitted bids on that date, with Booth and Flinn the low bidder at $19,250,000. This company was able to underbid its competitors because it already had a thirty-five-hundred-horsepower air compressor located in Lower Manhattan. This plant, necessary for supplying air to the underwater shields, had been built for use on the Fourteenth Street subway tunnel project and would be available for use at the Manhattan end of the project, thus saving a great deal of money.[31]

Within a few days, the staff engineers tabulated the bids and recommended to the commissions that they award the contracts to Booth and Flinn. Holland confidently predicted that the contracts would be signed within a week, that ground would be broken for construction of the New Jersey ventilation shafts early in March, and that traffic would flow through the tubes within three years. He was wrong. The latest delay in awarding Contract No. 2 had come to the attention of the New York and New Jersey state legislatures, which were then in session. Both bodies announced plans to investigate their respective commissions, ostensibly to find out why the project had fallen so far behind schedule. But for some legislators, the investigations would also serve as a means of addressing personal or political agendas, and on the outcome of these investigations the fate of the project now rested.

8

A Tempest in a Teapot

The whole thing is a tempest in a teapot. Neither the Jersey City Authorities nor Mr. Boyle can delay any longer the progress of this tunnel.
— New Jersey Interstate Bridge and Tunnel Commission Chairman Theodore Boettger, reacting to complaints about the "secret" groundbreaking ceremony held in Jersey City on May 31, 1922

IN JANUARY 1922, New York Assemblyman Russell B. Livermore of Yonkers introduced a bill to reorganize the New York State Bridge and Tunnel Commission, allegedly because its members had held up progress on the tunnel. He soon admitted, however, that some of his information concerning the need for the bill came from his father, a prominent New York lawyer who, Livermore said, was a "warm personal friend of Mr. Adams" and had acted as the latter's attorney in years gone by. "They've talked the matter over," Livermore admitted, "and I've probably received some of my information that way." After Adams was revealed as the man behind the bill, Livermore was encouraged to let it die, which he did. Adams's attempt to get his enemies tossed off the New York commission had failed.[1]

At the New Jersey State Capitol in Trenton, Adams was the focus of the investigation rather than the instigator. Republican General Assembly Majority Leader William W. Evans introduced a bill to replace the Democratic members of the New Jersey commission with Republicans. This would mean that Adams, Boyle, Shay, Stevens, and Gallagher would be tossed out. One Republican "high in authority," according to the *Jersey Journal*, stated, "We are specially determined to get at T. Albeus Adams, the president of the New Jersey Tunnel Board. We regard him as the main cause of the tunnel trouble."[2]

Hague, wishing to protect his good friend Boyle, and Adams, if he could, declared war on the Republicans. He refused to allow the Jersey City Board of Commissioners to approve the ordinances for closing the streets needed for the entrance and exit plazas. These ordinances had to be passed before the Erie Railroad agreement could be executed, thus allowing contractors to begin work in the railroad's yard. If the city did not voluntarily vacate and turn over the streets, the Erie Railroad would renege on its commitments, and the only way the land and easements could be obtained would be by condemnation. It was known that the Erie Railroad would attack the constitutionality of the condemnation act in court, and the process could drag on for years. Hague said, "Maybe they think they can bluff me. I wish them luck. If they have any idea they can put this thing over on us they have another guess coming. If they throw this commission out then we'll throw the whole tunnel plan into the discard."[3]

Adams repeated Hague's threat on February 23, 1922, when he submitted a revised annual report of the New Jersey commission to the legislature, covering changes since the official report had been printed on January 16. Along with the report, he sent a letter stating that if the bill was not withdrawn, the necessary Jersey City ordinances would not be passed. "If the ordinances are not passed, it means that the rights and privileges required on the Erie property will have to be secured by condemnation," Adams wrote. In that event, the Erie would start a long court fight, "and in the meantime," Adams warned, "the bonds outstanding and the ordinary expenses of the commission and its engineering force will pile up enormous costs for the people of New Jersey to pay."[4]

The New York commission refused to allow the award of Contract No. 2 until the ordinances were approved, out of fear that rapidly rising prices for labor and materials, combined with possible delays of unknown length, would render the contract price inadequate before construction could actually begin. Once again, politics, corruption, and greed had worked to delay the project. But Adams's day of reckoning was almost at hand.

The galleries of the New Jersey General Assembly chamber began to fill as early as 7 p.m. on the evening of March 6, 1922, in anticipation of a speech to be made by Majority Leader Evans. Democratic Assemblyman Henry J. Gaede of Hudson County, an ally of Hague, had demanded that Evans back up the charges that had been made against Adams. At about 11 p.m., Evans indicated he was ready to be heard. As the *New York Evening Post* reported, "An audience consisting of politicians from all over the State, which crowded the galleries and overflowed into the legislative corridors, rushed for places as Mr. Evans began his attack on Adams and his conduct of the commission. The chamber became as silent as a tomb."[5]

Even the Democratic members were quiet as Evans, his voice quivering with emotion, read from a *Post* reporter's affidavits, which recited the results of the newspaper's investigations. The audience did not remain silent for long. "Expressions of astonishment arose from all sides of the chamber as Mr. Evans recited, one after another, the purchases of land which companies Adams was interested in had made near the tunnel's mouth since he became a member of the New Jersey Commission," the *Post* reported. Adams, head bowed, eyes fixed on the floor, sat and fumbled with a cigar that was as dead as his future with the commission. Never lifting his head, Adams began to make copious notes as Evans warmed to the attack. The *Post* reported, "There was hardly a person in the garishly lighted chamber, including the newspapermen, who didn't breathe a sigh of relief when the tenseness and stillness caused by the recitation of the charges against Adams were broken. The only laugh came when Mr. Evans described the intricate land deals by which he said Mr. Adams' mother-in-law first bought property near the tunnel and then conveyed it to men associated in business with Mr. Adams."

Those deals, Evans charged, demonstrated that Adams was not a fit chairman of the commission. Evans also noted the way that Adams and his friends in Jersey City had conducted themselves in regard to the street-widening issue. Although Jersey City had pledged to vacate Eleventh Street in the interest of the tunnel project, there had been no action for two months. Moreover, "a reputable journal of Democratic persuasion" had informed its readers that the tunnel would never be built if the present commission was ousted because Jersey City would renege on its pledge to vacate the land.

"I will have this newspaper clipping entered as part of the record and read by the clerk," Evans said.

"I object," interrupted Mr. Gaede, "if it reflects on any member of the present commission."

"Oh, no," replied Mr. Evans with polite irony, "it reflects on no individual member of the commission. It reflects on the Democratic party of New Jersey."

The clerk then read the clipping, an editorial from the *Hudson Dispatch*. "The Republicans may be all powerful in the Legislature," read the editorial, "but Frank Hague is all powerful in Jersey City, and will be for some time." At this, according to the *Post*, the Democrats cheered and the Republicans hissed. Assemblyman William George of Hudson County then charged that Adams had been first appointed by the Republican Governor Edge, now a U.S. senator, and that Edge had twice reappointed the man the Republicans were now trying to oust. Assemblyman Arthur Pierson of Union, the "dean of the Assembly," who had been sitting quietly in his seat in the center of the chamber, leaped to his feet. Pulling off the green shade that protected his eyes from the glare of the chandeliers which dropped directly over his desk, he shouted, "Yes, Edge did. But let me say that Edge is more ashamed of that appointment than of any other single act in all his public life. Let me also recall that we all make mistakes."

Gaede interrupted Pierson to say that if the Republicans forced through the bill, which contained a provision granting the new commission authority to condemn the land needed by the project, the tunnel would be held up for years. "You can't go through with it," Gaede asserted. "If you want to commit political burglary, go ahead with this bill. You'll regret it next fall."

"The provision about the streets," shot back Mr. Evans, "is to keep Jersey City from carrying out its announced plan of blocking the tunnel. Jersey City plans to hold up this project. That's the need of the bill. We haven't acted hastily, but only after long consideration. We have seen this tunnel made the football of politics."

The Democrats continued to fight the bill until well past midnight, while Adams went from one assemblyman to another, trying to shove documents into the protesting members' hands. But the bill was passed to its third reading, by a vote of forty-one to eighteen. Later in the day, the Senate approved the bill by vote of fifteen to one. Edwards vetoed the bill, but it passed over his veto.

Adams, Shay, Gallagher, and Stevens were out. The members retained were Boyle, Noyes, Boettger, and Barlow. The new members were Isaac B. Ferris of Camden, Robert S. Sinclair of Newark, Judge John B. Kates of Camden, and Frank L. Suplee of Gloucester. It was also agreed that Edward I. Edwards, the governor's son, would be allowed to keep his $5,000-a-year job as treasurer of the commission. The retention of Boyle and Edwards was a trade-off, agreed to by the Republicans in order to win enough votes to get rid of Adams and the other Democrats. The Republicans also thought that by allowing Boyle to keep his appointment, they might obtain some softening of Hague's position on the street vacation. They did not.

On March 20, 1922, the newly appointed New Jersey commission held its first organizational meeting in Trenton, without the presence of Boyle, and elected Boettger chairman and Kates vice chairman. Kates offered a resolution for reinstatement of the dismissed Board of Consulting Engineers, with restitution of back pay. Wells and Clarke were officially dropped as technical advisors to the commission. The next day, the two state commissions held a joint meeting in New York and approved Kates's resolution, to become effective as soon as Wilgus withdrew his suit for back pay.[6]

The commissioners also voted to award Booth and Flinn the contracts for driving the underwater segments of the tunnel and constructing the New Jersey shafts. The company would have five days to sign the contract and thirty days to begin work. The commissions then appointed a committee to discuss with Hague the vacation of Eleventh Street in Jersey City and announced that they expected to obtain the vacation order within days. Hague immediately expressed his desire to proceed with the vacation, as soon as the Erie Railroad reached an agreement with the twenty property owners along Twelfth Street whose land was to be acquired by the railroad. Hague also resurrected the demand that the railroad maintain a permanent easement for a siding to the Union Terminal Cold Storage Company building.[7]

Early on the foggy morning of March 31, 1922, a fifteen-man work crew from Booth and Flinn gathered around Clifford Holland in the little triangular park at Canal Street where the New York groundbreaking ceremony had been held in October 1920. On either side of the park, the recently completed sixty-foot-deep shafts were planked over, awaiting the next stage of construction. Holland took a pick from the hands of a company laborer and drove it into the damp earth. General Superintendent Michael Quinn, of Booth and Flinn, then thrust a shovel into the loosened soil and threw it to one side. After watching several ceremonial swings, the laborer asked for his pick back, but Holland refused. The tool was plated with silver and hung in his office.

The excavation begun that morning was for the foundation that would support the huge thirty-five-hundred-horsepower air compressor, then sitting idle at Fourteenth Street and Avenue D, once used on the Fourteenth Street subway project. That plant and a similar compressor, yet to be built, would be used to pressurize the shields that would drive the underwater segments of the tunnel reaching out from each shore.

With work already under way in Manhattan, George H. Flinn became anxious for work to also begin in New Jersey, and within a few days, he sent a letter to Holland asking for immediate possession of the Jersey City work site. Holland took the matter to the commissions, and on April 4, the New Jersey commission turned up the heat on Hague by stating that if Jersey City did not vacate Eleventh Street within two weeks, they would condemn the land.[8] The Jersey City Commission, however, would not budge.

They decided that the price offered to property owners on Twelfth Street by the railroad was too low. According to the *New York Times*, "It has been intimated that this squabble was provoked by a Jersey City element that desires to harass the Tunnel Commission for the sole purpose of holding up the tunnel."[9]

Three weeks later, with no action taken by Hague or the Jersey City commissioners, the New Jersey tunnel commissioners decided to take a firmer stance. They let Hague know that they intended to abrogate the street-widening compromise reached the previous June. Henceforth, they would adopt the previous position of the New York tunnel commissioners and refuse to pay for any street widening beyond the entrance and exit plazas. Hague's response was to demand that the original agreement of August 1920 be honored, which would require the project to pay for $1 million in street widening.[10]

At this crucial point, a weakness in the strategy of the commissions became apparent. At a hearing in Jersey City the last week of April, Holland admitted under questioning by Jersey City Commissioner James F. Gannon that vacation of Eleventh Street was not actually necessary in order to build the tunnel. Although Holland later tried to deny that he had said any such thing, the truth was that the vacation was only necessary because of the deal that had been worked out with the Erie Railroad. In other words, the threatened condemnation was fatally flawed because the commission would not be able to prove in court that condemnation of the street was for a valid public purpose, as opposed to merely being an option desired by the railroad.[11]

The tunnel commissioners realized that they had to amend their plan of attack. They quickly hammered out a supplemental agreement with the railroad, signed May 28, allowing Booth and Flinn crews to occupy the work site in the Erie yard without first requiring the vacation of Eleventh Street. The railroad officers agreed to grant the easement because they knew that it was in the company's best interest for the project to move forward, and they realized that the commission had a weak position from which to force action by the City. A separate agreement was also reached with the Lackawanna Railroad covering a slight intrusion onto its property. The new agreements not only allowed construction to proceed; they also provided Holland, the Board, and the New York commissioners with a means of wrecking the New Jersey commissioners' plans for a grand groundbreaking on their side of the river.

Early in the afternoon of Wednesday, May 31, 1922, a group of about twelve men dressed in suits, one carrying a pick, one a shovel, one a pry bar, and another a camera on a tripod, stepped off a ferry in Jersey City and began slinking north toward the Erie Railroad yard. They were careful to avoid the streets and tried hard not to be seen. But the strange sight of such elegantly attired gentlemen walking with laborer's tools brought the attention of various police officers, who stopped the group several times before allowing them to proceed.

About a mile away at City Hall, the Jersey City Commission met to adopt a resolution commending Hague and Boyle for holding up construction of the tunnel until they got their way in the matter of street improvements. Holland, however, had decided that it would be hard to stop work that had already begun. It was he who led the small cluster of men as they approached a spot in the Erie yard within rock-throwing distance of the Union Terminal Cold Storage building. With Holland were two of his staff, Construction Engineer Milton Freeman and Resident Engineer M. I. Kilmer. Five representatives of the contractor, Booth and Flinn, were also there: George H. Flinn, William A. Flinn, General Superintendent Michael Quinn, Superintendent Le Roy Tallman, and Chief Engineer M. E. Chamberlain. Two officers of the Erie Railroad, Chief Engineer R. C. Falconer and Special Engineer H. F. King, were part of the group. It was they who had represented the railroad in all the negotiations for easements and land acquisition. A photographer from the *Engineering News-Record* and a reporter from the *New York Times* followed along to record the clandestine ceremony that was about to occur.

Holland took a shovel, George Flinn a pick, and Falconer a "pinch" bar as they posed between two rails and let the photographer expose a few plates. Falconer then pried up a section of track so that Holland and Flinn, both with straw hats set firmly on their heads, could go to work digging a shallow hole. With the deed accomplished, the group returned to the ferry landing as surreptitiously as they had come and were soon were back in Manhattan.

When asked by the *Times* reporter what the Jersey City officials would do now, Holland replied that he did not know. As soon as the shafts were sunk, the work of driving the shields would begin, and three and a half years later, the tunnel would be complete. "The tunnel commissioners may again take up the matter of street improvements in Jersey City if there is anything left of the subject by that time," Holland added.[12]

When Hague found out about the ceremonial act, he was so upset that he could not face newspaper reporters, leaving it to a subordinate to state,

"The action of the tunnel officials in breaking ground in secret was just what might have been expected from the tunnel commission that refused to provide for safe and adequate approaches on the New Jersey side of the Hudson, while planning to spend $1,500,000 of New Jersey money for safe and imposing approaches on the New York side."[13] The figures provided by Hague were based on the estimated costs of acquiring sixty-two parcels of land in Manhattan lying in five city blocks between Varick and Hudson Streets, and from Laight Street on the south to Spring Street on the north. Although it had originally been assumed that almost all the land needed for tunnel entrance and exit plazas in New York would be acquired from New York City, new plans which included the open-cut areas resulted in the need to acquire more private property. These properties were considerably more valuable than land in Jersey City because they could be used for purposes other than industrial. The New York commission was not, however, improving New York City streets beyond the plazas, as Hague demanded they do in Jersey City.[14]

Jersey City Corporation Counsel Thomas J. Brogan told Hague there was not much that the city could do to stop the work as long as it was on Erie Railroad property. Within a few days, however, Hague thought of something that could be done. By nature, Hague was a brawler. As a young boy growing up in the Horseshoe slum of Jersey City, he learned that a fist to the face was often an effective way of settling arguments. As a teenager, he briefly trained to become a prizefighter, before coming to the realization that there was more money to be made as a politician. He never lost his tendency toward physical violence, however, and while mayor of Jersey City, he was known to physically assault those who displeased him, including police officers, fire fighters, and other city employees. On one occasion, after witnessing a fire in which a person was injured and telephoning for an ambulance, he was enraged when it took forty-five minutes for the vehicle to arrive. Hague asked the doctor who arrived with the ambulance why it had taken so long. Not liking the response he received, Hague hauled off and smashed the physician in the face.[15] The bare-knuckle style of New Jersey politics suited Hague's temperament, and he was not above using any means, fair or foul, to win a contest. In his dispute with the New Yorkers of the tunnel commission, he had one more punch to throw.

Sinking of the New Jersey shafts began in earnest on June 15, and soon thereafter, crews from Booth and Flinn began erecting a powerhouse at Twelfth and Provost Streets. They had just begun their work when Jersey City police and the city building superintendent showed up and asked to see their building permit. The crews did not have one, so the police ordered them, under threat of arrest, to stop work. Hague had once again held up the project.

Two weeks later, New Jersey State Chancellor Edwin Walker issued a temporary restraining order preventing Jersey City from interfering with the tunnel project, but no one knew what his final ruling might be. Holland, fed up with all the political interference and delays, threatened to quit if the injunction was not made permanent. He need not have worried. On July 13, Walker ruled that the work was being conducted by authority of the state in its sovereign capacity, and thus the entire project was beyond the control of Jersey City. Brogan, knowing that the city could not win, announced that he would not appeal the decision. The fight was over.[16]

Another lingering issue also came to an end during this period. Although the newly appointed New Jersey commission had voted to rescind the dismissal of the Board of Consulting Engineers, it was unclear what the Board's continuing role on the project would be. When the commissions created the Board in 1919, it was assumed that its services would be required until the plans for the tunnel were complete and adopted and the main portions of the work under contract. By the end of March 1922, both of these events had occurred. The commissions therefore voted to dissolve the Board, effective May 31, 1922. Each Board member would be offered the opportunity to stay on as a consultant to Holland, on an as-needed basis at the rate of $150 a day. There was little left for them to do, however, and the offer was more of a courtesy than an expression of need for their services.

With most of the major design decisions made, contracts signed covering all the shaft and tunnel excavation, and a new spirit of harmony existing between the state commissions, Holland and his staff could now concentrate on the task that they had waited so many years to accomplish—the construction of the tunnel. That work began with sinking of the ventilation shafts on each side of the river and with construction of the tunnel-boring shields that would be used to excavate the tunnel tubes. There would be a total of seven ventilation shafts and six shields in operation.

In Jersey City, there would be a land shaft and a river shaft for each tunnel, making four on the New Jersey side of the river. One shield would be driven eastward from each land shaft through the river shafts and onward to a rendezvous with the shield being driven from New York. Another shield would be driven westward from each land shaft, underneath the yards of the Erie and Lackawanna Railroads, to the point where the tunnel would intersect with the open-cut approach, with one approach leading to the exit plaza and the other leading from the entrance plaza. Thus, there would be four shields in use on the New Jersey side.

In New York, there would be a land shaft for each tunnel, but only one shared ventilation shaft. This was possible because of the close proximity of the tunnels at the New York pierhead line, which was not possible to achieve on the New Jersey side of the river. A shield would be driven westward from each New York land shaft through the river shaft and onward to a rendezvous with the shield coming from New Jersey. It would not be necessary to drive shields eastward from the New York land shafts because there was no railroad property to protect. The tunnels could be constructed all the way to the shafts, either in publicly owned streets or through property acquired for the project, using the cut-and-cover method. This involved excavating all earth down to final grade, constructing the approach tunnels, and then covering the tunnels and rebuilding the streets.

On August 8, 1922, the commissioners and engineers went on an inspection trip, beginning at the Canal Street shaft, where excavation had been completed a year before, on August 9, 1921. They next looked at the Spring Street shaft, about two hundred feet away, where excavation had been completed July 15, 1921.[17] Acceptance of the shafts had been delayed until December 1921, due to the need to shift certain surface and subsurface structures, including the tracks of the New York Central Railroad. This work had to be performed under a separate contract.[18]

The engineers sank each shaft using a caisson, a huge, double-walled, rectangular steel box, open at top and bottom, with an air-tight working chamber in the lower portion formed by a reinforced horizontal partition about seven feet up from the bottom of the walls. This partition formed the "roof" of the air chamber. The bottom end of the walls formed a cutting edge which dug into the earth. Hoses passing through each chamber's roof supplied air to pressurize the interior, thus keeping water out as the caisson descended below the water table. As the caisson descended, the amount of pressure had to be increased to keep out the water.

Each chamber's roof was also pierced by two air locks through which material excavated from within the air chamber could be passed. This material, mainly soil and rocks but also old pilings and fill, was temporarily dumped onto the top of the chamber as the caisson descended, thus helping force the caisson down. The walls of the caisson were also filled with concrete as it descended, thus increasing the weight and forcing the structure farther down. Between each air lock reserved for material excavation there was another T-head lock, twenty feet long, four and a half feet high,

and three feet wide, where work crews could pass through. In each T-head
lock, there were two wood benches, one along each side, onto which
twelve to fifteen sandhogs, depending on the size of the crew of a particular
shift, would sit while waiting to pass into or out of the air chamber. Each
T-head lock also had an air-pressure gauge and a clock, so that an operator could keep track of the amount of pressure and the time workers spent
decompressing.[19]

Once each caisson reached the desired depth, workers sealed the bottom at the cutting edges, removed the air locks, and filled the air chamber with concrete. The roof of each air chamber then became the floor of
a working space in which the shields would be erected. As each caisson
descended, temporary circular bulkheads were constructed in the east and
west walls. Once the shields were assembled and the process of tunneling
ready to begin, a new roof would be constructed above the shield to create a new air chamber. This chamber would then be pressurized and the
west bulkheads removed so that the shields could be driven toward the
river. The east bulkheads would likewise be removed to accommodate
the approach sections, which would be excavated by an open cut from
the surface.

The contractor's power plant provided both low-pressure air for the
working chambers and high-pressure air for caulking, riveting, hoisting engines, and other work in the caissons. Later, high-pressure air would also
be needed in the tunnels for the hydraulic jacks and erector arms of the
shield. Three boilers, with a combined capacity of 350 horsepower, were
also provided to supply steam.

As the Canal Street caisson descended, it broke through the cribbing
(support walls) of an old dock. Prior to 1800, Greenwich Street was the
riverside drive leading from the lower part of the city to the more rural area
to the north. All the land west of Greenwich Street had been reclaimed
from the river over the years. When New Yorkers built the dock, the shoreline of the Hudson River was about 250 feet farther to the east than its
present position.

At the time of the inspection trip, the Canal Street shield, which would
be used to drive the south tunnel, was still under construction by the Merchant Shipbuilding Company of Chester, Pennsylvania. After completion,
it would be assembled at the plant to make certain that everything fit, then
disassembled and shipped to New York for reassembly within the shaft.
Delivery was anticipated on September 1. Within a few weeks, the shield
would be ready to begin its journey toward the point in the river where it

would meet up with another shield driven from New Jersey. If everything went well with the Canal Street shield, the Spring Street shield would be assembled a few months later.

After examining the New York land shafts, the engineers and commissioners crossed the river and debarked at Pier 9 in the Erie Railroad yard, not far from the power-plant site on that side of the river. Due to the delay in shaft excavation in Jersey City, the caissons there were still under construction. The commissioners were happy, nonetheless, to see that work was finally progressing.[20]

At about 9:25 on the morning of October 26, 1922, engineers supplied air pressure to the hydraulic jacks of the Canal Street shield, and it began its slow movement toward the river. The shield was thirty feet, two inches in outside diameter, and sixteen feet, four inches in length. The upper half beyond the cutting edge formed a hood that projected another two feet, six inches from the front of the shield. Five vertical and three horizontal walls divided the shield into thirteen working compartments, through which the sandhogs could access the earth in front of the shield. Thirty hydraulic jacks were spaced evenly around the back of the shield, and together they had a combined thrusting force of six thousand tons. A counterweighted erector arm attached to the back of the shield on a central pivot shaft was used to grab the cast-iron plates forming the outer tunnel wall and place them into position for bolting. Each ring segment required 160 bolts, weighing ten pounds each. The entire shield, including equipment, weighed four hundred tons. It was the largest of its type ever constructed.

The ventilation studies had led to an increase in the outside diameter of the tunnel from twenty-nine feet to twenty-nine feet, six inches. This small increase accommodated larger air ducts for maximum efficiency and economy of air flow. Since the shield had a larger diameter than the tunnel, it actually overlapped the tunnel at its tail. The void between the cast-iron plates and the shield was filled by a special grouting mixture injected through a nipple in the tunnel walls by a pump. As the sandhogs completed each ring segment, they grouted the void beyond the tunnel walls and then shoved the shield forward.

The forward movement of the shield was not a continuous movement but an endless series of small shoves of two and a half feet each. This was the width of an individual ring, with each ring consisting of fourteen plates, each plate being six feet long and one and seven-eighths of an inch thick and weighing thirty-three hundred pounds. Each ring also included a "key" segment weighing about twelve hundred pounds. The completed rings formed a surface against which the jacks would exert pressure. The direction of shield movement was controlled by applying differential pressure to the jacks. If the tunnel needed to go slightly to the left to stay on course, more pressure would be applied by the jacks on the right, and vice versa.

At the front of the shield, sandhogs excavated a pocket at the top of the working area in advance of the hood and packed it with clay brought from Staten Island. The hood then pushed through this mass of clay as the shield advanced. This formed a seal that helped prevent losing air upward through the soil. A fixed, wood bulkhead was placed in front of the bottom quarter of the shield, and as the shield moved forward, it pushed against this bulkhead, forcing the soil and rock through openings in the bulkhead or over the top of the bulkhead into one of the working compartments where it could be shoveled out.

As the shield moved forward, a worker sitting at a monitoring station paid careful attention to the amount of air pressure in the working chamber. If the pressure was not great enough, the soil and water could rush into the working chamber faster than it could be removed. If the pressure was too great, it could force the air out of the working chamber, through the soil and up to the surface at such a rate that there would be a "blowout" of air and earth. Once the shield was under the river, a blowout could result in an uncontrolled inrush of water that would fill the working chamber very quickly, drowning the sandhogs before they could escape. It had happened before.

A major blowout on the first tunnel under the Hudson River, one of the Hudson and Manhattan tubes, caused twenty men to drown in July 1880. The Battery-Joralemon Tunnel, the first project that Holland had worked on, also had blowouts. In one particularly spectacular incident, on March 27, 1905, a rapid outrush of air caught a sandhog and blew him out of the tunnel, up through about thirty feet of river mud and water, and left him floating, alive, in the East River. Another incident on February 19, 1916, in one of the Whitehall-Montague tubes, resulted in three men being blown upward through twelve feet of riverbed, through the river itself, and then hurled atop a geyser twenty-five feet in the air. Only one of those men, Marshal Navey, survived. Rescuers found him swimming about somewhat aimlessly in the river and took him ashore in a skiff. The experience did not dissuade him from doing compressed-air work, however, and he was one of the sandhogs who worked on the Hudson River vehicular tunnel.[21]

The pressure needed to achieve the optimum balance between not enough and too much depended on a number of factors, including the shield's depth and the type of material being driven through. Much of the material under the river would be a thick, muddy mixture of silt and water referred to as "muck." This material often streamed through the openings in the shield face into the working chamber in long "sausages" that the sandhogs could break up and load into carts. In other areas, the sandhogs would encounter solid rock that had to be blasted with explosives to reduce it to sizes small enough to pass through the openings in the shield.

Between October 26 and December 2, 1922, the Canal Street shield progressed far enough to allow erection of twenty-seven tunnel rings forming nearly sixty-eight feet of completed tube. But as the shield passed within five feet of the cofferdam (wall) enclosing the excavated area for New York City's new sewage-treatment plant, air leaking from the shield blew about 150 cubic yards of earth into the excavation. This not only imperiled the sewage-treatment plant site, but it also left cavities in the earth that could have led to a blowout. Work in the tunnel had to be held up while workers constructed the permanent concrete walls of the plant.

Before commencement of tunneling through the river bulkhead, crews deposited large amounts of clay between Piers 34 and 35 and on the landward side of the bulkhead to help prevent a blowout. Great care needed to be taken, because if the piles of the old bulkhead were not removed in advance of the shield's approach, the shield could push against them, thus creating a dangerous void. As many as thirty piles had to be removed at one time to advance the shield the length of one ring. The engineers expected that driving the shield would be much easier once it advanced beyond the bulkhead.

On December 5, as work continued on the sewage-treatment plant, a long-anticipated event took place in the Staten Island shipbuilding yard at Mariner's Harbor. Ann H. Holland, the chief engineer's twelve-year-old daughter, broke a bottle of champagne against the double-walled caisson that would form part of the New York river ventilation shaft. As she completed her swing, workers knocked away the blocks holding the caisson, which slid down greased rails and splashed into the water. Measuring thirty-five feet high, ninety-three feet, three inches long, and thirty-seven feet, three inches wide, the great steel and concrete structure, the largest of its type ever built, tilted back and forth a few times before stabilizing. A tugboat pulled up alongside and took the caisson in tow, moving it to a nearby dry dock where concrete would be added to make it fifteen feet taller. When ready, it would be towed to a spot near the end of Pier 35 and guided onto a pile-supported platform, open on the south end, where additional steel would be added to the walls and more concrete poured between the inner and outer walls.

Workers assembled the Spring Street shield, identical in all dimensions to the Canal Street shield, on the last day of 1922. Pressure was applied to the shield chamber on January 8, 1923, and on January 17, the shield began moving westward. Both shields would meet at the New York river ventilation shaft at the end of Pier 35.

On the New Jersey side of the river, the north land caisson had descended only a fairly short distance into the cinder fill of the railroad yard when crews encountered a rock-filled timber crib on November 15, 1922. The timbers had to be sawed or chopped into short lengths and then brought up to the surface, along with the rock. Workers encountered a similar problem in the south land caisson on December 21, 1922. By New Year's Day 1923, the north caisson had been sunk only about thirty-eight feet and the south caisson about nineteen feet.

Another problem slowing work in Jersey City was the depth to bedrock on that side of the river: 250 feet, as opposed to just 70 feet on the New York side. This was far below the depth to which sandhogs could work, due to the excessive air pressure required to keep water out of the caissons. Therefore, the river shafts could not be sunk all the way to bedrock. Because the silt which overlay bedrock in that area would not provide a stable support for the two river shafts, the engineers developed a plan to transfer the 10,155-ton weight of each river shaft to forty-two piles, extending from the bottom of the river shaft down to bedrock. Each pile was composed of a steel cylinder twenty-four inches in diameter, filled with concrete. The first pile was driven on April 14, 1922, and all of them were in place by March 1, 1923.

While this work was going on in New Jersey, great progress was being made in New York. At about 5:45 in the morning of January 31, 1923, the massive 1,650-ton New York river-shaft caisson began its slow journey from Staten Island, across the bay, and up the Hudson River. With recent extension of its walls, the structure now rose thirty-five feet above the water line and extended twenty-five feet below it. The trip was planned to take advantage of a favorable tide, but a stiff wind from the north made it difficult for the three tugs towing the caisson to make headway. At one point, they had to anchor near Bedloe's Island to keep from being pulled back the way they had come. Two Booth and Flinn employees riding on top of the structure became so cold that they built a fire and huddled around it. The caisson did not arrive at the end of Pier 35 until around 8 p.m., when it was too dark to guide it within the piles prepared for it.[22]

The next morning, the caisson was secured in position and the work begun of building up the steel walls and pumping concrete into the five-foot space between the inner and outer walls. As the weight increased, the caisson sank about thirty-five feet, at which depth the cutting edge reached the riverbed. Compressed-air operations began as soon as the caisson settled firmly in the silt, and the excavation proceeded at a rate of 7.5 feet in a twenty-four-hour day when going through silt and 6.2 feet per day through sand. When rock was encountered, at a depth of about 69 feet below mean high water, the rate of descent slowed to 2.6 feet per day. The caisson would not reach its final position, ready to receive the shields, until the end of the year.

The much smaller New Jersey north river caisson launched on January 3, 1923, but was not towed into position until March 30. The identical south river caisson launched April 2, 1923, and arrived at the shaft site on August 24, 1923.

Sinking shafts, constructing and sinking caissons, and driving shields through the earth presented challenges never before encountered in subaqueous tunnel construction, due to the unprecedented size of the tunnels, the overall scale of the project, and the peculiarities of the location. As just one example, after the south tunnel from Manhattan passed the river bulkhead, it tended to shift position more than engineers had predicted. They expected that the river's tidal action would move the tunnel a few inches this way or that, but it was moving too much. On a hunch, Holland had a sample of water from the bottom of the river tested. It was fresh. Water taken from that portion of the river should have been salty. This led to further investigation that revealed a spring below the riverbed, fed from an underground stream. This was the same stream that once fed the old Collect Pond. The stream's water supply was so great that it affected the stability of earth far below the riverbed, thus allowing the tunnel to shift more than it should have. Once the engineers identified the cause of the problem, they were able to solve it by blocking off the flow of water.

Other challenges unique to the project would be solved as they arose, as would all the myriad routine problems of subaqueous tunnel construction with which Holland and his staff were familiar from their work on subway projects. In one respect, however, they hoped that past experience would not be repeated. They were determined that the injuries and deaths common in tunnel construction be reduced as much as possible. They also wanted to avoid the labor strikes that had so often stopped work on other projects. By the beginning of 1923, however, accidents, injuries, and deaths had already occurred. There would be many more. And the laborers would strike.

9

The Sandhogs

As word of the bizarre world under the river became known to the public, people envisioned a race of superhuman men who were able to work under impossible conditions. But the sandhogs were very human.

—Paul E. Delaney, *Sandhogs: A History of the Tunnel Workers of New York*

THEY WERE, MOST of them, immigrants. In the nineteenth century, they came from England, Scotland, Germany, and Italy, although the majority came from Ireland. Beginning in the first decade of the twentieth century, some of them were from Austria, Poland, Hungary, or the West Indies. The work they did was dangerous, even by the low worker-safety standards of the age in which they lived. Some of the nation's greatest works of engineering, particularly its largest bridges, tunnels, and buildings, could not have been built without them.

A letter writer to the *New York Times* in April 1923, reacting to an article about a recent strike by workers in the south tube of the Hudson River vehicular tunnel, suggested that they be called "Pressure Men" because they worked in compressed-air environments. He found this term to be far more respectful than that by which they were commonly known: "sandhogs." "There is nothing hoggish about them," he asserted, "as I can testify from an experience extending over half a century."[1] The writer, perhaps a sandhog himself, may well have traced his experience back to May 8, 1872, when compressed-air workers first struck to protest hazardous working conditions in the New York caisson of the Brooklyn Bridge. The strike came after several workers had died from decompression sickness, later called caisson disease and commonly known as "the bends."

The popular term for the disease came from St. Louis, where workers in a caisson of James Eads's great steel-arch bridge over the Mississippi River experienced the ailment in 1870. Coming to the surface at the end of a shift, some men began to feel a muscular paralysis of the lower limbs. As the project continued, the caisson descended to greater depths and the air pressure within the working compartment increased. With greater pressure came severe headaches, unbearable pain in the joints, or acute abdominal cramps. Men experiencing this condition sometimes walked about slightly bent over. This posture was referred to by the sandhogs as the "Grecian bend." The workers thus began referring to the affliction as, simply, "the bends."[2]

The symptoms of caisson disease were caused by a too-rapid decompression of the worker's atmospheric environment. Under high pressure, inert gasses, particularly nitrogen, dissolve in the blood. The greater the pressure, the more nitrogen dissolved. The nitrogen is harmless as long as it stays dissolved. But if the pressure is decreased too rapidly, the dissolved nitrogen can come out of solution and form tiny bubbles that lodge in blood vessels and soft tissues, causing severe pain, long-term disability, or death. This is what happened to the compressed-air workers in the caissons of the Eads Bridge, the Brooklyn Bridge, and in all the tunnels built under the Hudson River and East River prior to construction of the Hudson River vehicular tunnel.

It is hard to know just how many men died or suffered severe injury as a result of caisson disease while working underground in New York. Contractors tended to place little value on the lives of their workers, who could be replaced easily, and it was to the contractors' advantage to skimp on safety. It was cheaper to allow a certain number of deaths or injuries to occur than it was to provide a safe working environment. In an attempt to limit liability, and to avoid public or governmental calls for improved safety procedures, contractors routinely blamed workers for accidents and encouraged falsification of death certificates. Sandhogs' deaths were often attributed to "natural causes," even when it was known that they died from the bends. Records were also inaccurate because in many cases men were fired after becoming ill and sent home to die or to suffer for years from the lingering effects of the disease. Merely complaining about working conditions could get a man fired, and there were always new applicants willing to fill a vacancy.

Worker safety in the tunnels was, to a large extent, an "economic externality," not fully part of the pricing equation when contractors bid jobs. And contractors fought stubbornly against improved safety whenever they thought that their profit margin might be in jeopardy. One example is S. Pearson and Sons, contractors of the Pennsylvania Railroad tunnels under the East River. The company replaced skilled air-lock tenders early in 1906 with inexperienced men in order to reduce cost by a few dollars a day, even though skilled tenders were essential to safe operations. After newspapers published charges that the company had been hiding deaths and injuries, the families of victims forced an investigation by New York County Coroner George F. Shrady that began on June 5, 1906. Witnesses testified that more than one hundred men suffering from the bends had been removed from the tunnel since the first of the year. They also stated that medical inspectors allowed men to work who were physically unfit; that the company had tampered with air-pressure gauges so that workers would not know how much pressure they were subjected to; that there were no toilets in the tunnels and the men had to relieve themselves where they worked; and that there was not a single hospital in New York equipped to treat caisson disease.[3]

At about 6:15 a.m. on June 20, 1906, a blowout in the southernmost East River tube drowned two sandhogs, one West Indian and the other Polish, and injured several other men. A representative of S. Pearson and Sons claimed that this type of accident was inevitable in compressed-air work. A few hours later, the two-week-long coroner's investigation resulted in a "sever censure" of the company, and not much else.[4]

The company representative's claim was not true. There was only one recorded death from caisson disease in the Pennsylvania Railroad tunnels under the Hudson River, partly because the men in that tunnel were working under an average pressure of about twenty-six pounds per square inch, whereas the average pressure in the Pennsylvania Railroad East River tunnels was about thirty-five pounds per square inch. Yet the average pressure in the East River subway tunnels dug by the Rapid Transit Commission (later Public Service Commission) was also thirty-five pounds, and only two deaths were confirmed in all these tunnels due to caisson disease. The best documented evidence available indicates that there were 3,692 cases of caisson disease in the Pennsylvania Railroad East River tunnels and twenty deaths, 550 cases in the Pennsylvania Railroad Hudson River tunnels and one death, and 680 cases in the East River subway tunnels with two deaths.[5]

Unionization and strikes were the only tools available to tunnel workers to combat contractor exploitation, and union advocates often asserted then, as they still do today, that these tools brought about better working conditions. Perhaps they did. But when the workers struck, their demands often focused on better pay, which did nothing to protect their health.

Moreover, the historical record indicates that unionization and strikes were only partially effective in bringing about real improvement and that progress was slow and incremental. The greatest contribution that unions made to improved worker safety was probably the effect their strikes had in raising awareness among the public, politicians, and insurance companies.

Late in 1921, the Industrial Board of the New York State Department of Labor opened hearings in Buffalo, Rochester, Syracuse, and New York City regarding proposed rules relating to tunnel construction and work in compressed air. These hearings resulted in changes to the State of New York Industrial Code in 1922 that established the maximum number of hours allowed in atmospheres of varying pressure, permissible periods of decompression, provision of sanitary facilities, and provision of medical care. The regulations helped make the Hudson River vehicular-tunnel project much safer than previous projects, but the job of a sandhog, or that of any laborer on the project, was still dangerous and very hard. And any man who took exception to the risk or working conditions would likely be dismissed. There were many unskilled immigrants looking for work, and there would always be a replacement waiting to take the wary man's place. As one sandhog recalled, "The turnover in workers was unbelievable. Men would work an hour or maybe a shift and they'd never be seen on the job again. Even the strongest men were tired after fifteen or twenty minutes in the air. And there was always the worry of being fired. If a man went for more than two sips of water during a shift, he was told to collect his wages and go home."[6]

New applicants for a position as sandhog were examined by medical personnel to determine if they were physically fit enough to withstand work in a compressed-air environment. If hired, they reported to a building near the entrance which housed the engineers' field office, dressing rooms, showers, and a small emergency hospital. After donning the appropriate overalls, hats, and boots, the workers took a hoist or "cage" down to the bottom of the caisson. After walking away from the caisson and down the tunnel, they eventually encountered the huge concrete bulkhead that formed the rear of the compressed-air working chamber. This bulkhead was divided into two levels by a horizontal platform. On the lower level there were two "muck-locks," pierced by rails, through which excavated materials were brought out or supplies taken in. Climbing up the ladder to the second level, the men saw two "man-locks" extending through the bulkhead. One was usually reserved for emergencies. The other, used for normal movement of workers into and out of the air chamber, was about six feet in diameter and twenty feet long, with small doors at both ends.

Stationed just outside the air locks, a checker recorded how many men went into the working chamber.

Decompression of the air lock created a roaring noise, like a tremendous amount of steam being released. After a few minutes, the door of the lock opened, and workers from the previous shift stepped out, stooping as they exited. As they left the working chamber, the checker recorded how many came out. The length of time it took to decompress depended on how long and under what pressure the men had been working. After working at maximum pressure, they would have to spend as long as an hour in the lock before they could leave.

Then the new shift filed in and took places on the full-length benches that lined both sides. A lock tender pulled the door shut with a metallic clag and opened the valve allowing air to hiss into the lock. All eyes were on the gauge located above the opposite door as the needle moved, indicating the increase in pressure. The effects were felt immediately as the air grew warmer and more humid, and some men began to sweat. The reverse was true on the way back out, as the expanding air became very cold. But exiting workers benefited from heating elements in the lock.

To avoid injury or discomfort, men undergoing compression worked their jaws or held their noses and "blew" to equalize the pressure. The noise of inrushing air was so loud that it was difficult to hear, so if anyone could not equalize, he had to hold up his hand so that the rate of compression could be slowed. Sometimes men could not equalize or were so frightened by the process that they asked to be let back out.

After about ten minutes, when the pressure in the lock reached that of the working chamber, the lock tender opened the door at the other end, and the men stepped out onto an elevated platform that ran the full length of the tunnel along one side. Below, in the dimly lighted tunnel, one could see the two sets of muck-car tracks fading into the distance toward the shield. The sandhogs walked down a flight of stairs to the space between the tracks and started walking toward the shield. When explosives were used to blast away rock, the air would be thick with lingering gray smoke, which had nowhere to go. Even when there had not been recent blasting, the air was muggy and oppressive, with a strong smell of damp earth.

This portion of the tunnel usually looked deserted because all the work was occurring at the other end of the chamber near the shield. As the men walked in that direction, they saw surveyor's platforms suspended at intervals from the top of the tunnel. This is where measurements were taken as the tunnel progressed, to make certain that it was being driven in the right direction. The men also saw pipes running the length of the tunnel, supply-ing water to the hydraulic jacks of the shield up ahead. As they neared the working area, they saw above them a metal safety curtain stretching across the upper two-fifths of the tunnel. This shield acted as a water stop in case of a blowout, and once past it the escaping sandhogs could run down an elevated walkway along the side and top half of the tunnel, safe above the water level. Or so it was hoped.

The shift on duty could be heard before seen, and as the relief-shift workers walked up to the shield, the deafening clatter of pneumatic tools or the throb of the grouting machine assaulted their ears. The area just behind the shield was brightly illuminated by huge work lights, which re-vealed up to fifty laborers going about their tasks, along with an inspector and the engineer in charge of that shift. Some men would be up in front of the shield, scooping muck over the top of the wood bulkhead or ham-mering away at rock, depending on the conditions encountered at any one time. These men were called "miners." Others might be cutting the muck sausages pushed through the shield into blocks and placing those blocks into the muck-cars or shoveling rock into the cars. These workers were referred to as "muckers." Some others, called "iron men," would be wield-ing huge seventy-five-pound wrenches, used to tighten ten-pound nuts onto the nine-inch-long bolts holding the tunnel rings together. Still oth-ers would be connecting or disconnecting hoses to the grout nipples in the walls of the shield or cramming lengths of hemp soaked in red lead into recesses in the tunnel segments. The hemp helped seal the spaces between segments. Bags of sawdust lay open on the ground, and from time to time the men dipped their hands into them in order to secure a better grip on their tools. Mud was in and on everything, making tools hard to hold and the ground slippery.

New workers were often amazed to find that they could not whistle in a compressed-air environment, no matter how hard they tried. They were also told that if a fire started in the working chamber, it could not be put out, so they were not allowed to smoke. But there was no time for whis-tling or smoking anyway, as the sandhogs were pushed to work as quickly as possible. Their only break came when there was a "shove" of the shield. When this occurred, the shift engineer ordered the air pressure to be low-ered, after which the sandhogs opened the doors of the bulkhead so that the muck could be pushed through as the shield advanced. The men laid down their tools and stood back away from the shield as the damp air quickly became so foggy that a worker could not see another standing a few feet away. As operators opened valves controlling hydraulic pressure, the sandhogs heard the sound of water running through the pipes to the jacks as the shield screeched forward. Muck sausages oozed through the openings in the bulkhead, and some men cut off chunks of muck with shovels while others threw sawdust on the resulting blocks so that they could be picked up and put into carts. In instances when the muck needed to be retained in the forward part of the tunnel so that the added weight would help keep the tunnel from rising, the blocks of earth would be taken back about fifteen yards away from the shield and dumped.

As the shield moved, sandhogs stood at either side of the tunnel and measured the advance. This would be done not only to determine how far the shield needed to go to allow installation of a new tunnel ring but also to check the alignment. When the shield had advanced far enough, operators reduced the hydraulic pressure to the jacks, the air pressure in the chamber increased, and the erection of ring segments continued.

The work was exhausting, dirty, and dangerous, and sandhogs were paid well, relative to wages paid above ground, to do it. The miners and their helpers received seven dollars for an eight-hour day (almost eighty-seven dollars in 2010 dollars), the iron men and helpers six dollars a day, and the muckers five dollars a day. That was considered good pay for the time, about 50 percent higher than the wage rates paid to non-compressed-air laborers on the job. But how much pay was enough to compensate for the risks involved and the toll that such work took on the body and mind?

Up to January 1, 1923, there had been only four cases of caisson disease on the Hudson River tunnel project and no deaths from that cause. There had been deaths, however. On October 30, 1919, Philip Healy of the P. J. Healy Company, the contractor for river and land borings, drowned in the river about five hundred feet west of Pier 35. His was the first death associated with the tunnel. On August 21, 1922, Steve Roizek, an employee of Booth and Flinn, fell from a scaffold in the New Jersey powerhouse and became the second man killed. A falling gantry crushed another Booth and Flinn employee, Christopher Kelly, on November 2, 1922, and the company lost another man on December 19 when a flywheel in the New Jersey powerhouse struck John Hues. On February 18, 1923, Joseph Richard, another company employee, fell to his death in the Spring Street shaft. None of these men was working under compressed air.[7]

As the south tunnel being driven from New York went deeper and the air pressure increased, the number of reported illnesses due to caisson disease began to climb, and the sandhogs became agitated. On Saturday,

April 7, 1923, approximately 150 of them laid down their tools and walked off the job. They wanted an increase in pay to eight dollars per day for miners and helpers, seven dollars for iron men and helpers, and six dollars and fifty cents for muckers. They also wanted a decrease in tunnel pressure from twenty-one to eighteen pounds per square inch, which would mean shorter shifts. Under state law, a normal workday consisted of two four-hour shifts when the pressure was less than twenty-one pounds. At this pressure, few incidences of the bends were reported, so it was assumed that the men could be worked for eight hours. At pressures from twenty-two to thirty pounds, shifts were divided into two three-hour periods, with a rest interval of three hours. Under this pressure, the number of reported illnesses began to rise. For pressures from thirty-five to forty pounds, the sandhogs worked for one and a half hours, rested for three hours, and then worked for another one and a half hours.[8]

In addition, the strikers wanted recognition of their union. The Compressed Air Workers' Union and the Foundation Workers' Union merged in 1900, and the American Federation of Labor chartered the united group in 1904. In 1918, the Compressed Air and Foundation Workers' Union merged with the International Hod Carriers, Building and Common Laborers' Union of America, retaining the name of the first organization. In 1920, the union backed efforts of African Americans to win equal treatment by other unions, and there were many African Americans employed on the Hudson River tunnel project. They often worked at certain defined tasks within a crew, particularly iron work, and were integral members of the workforce.

A representative of Booth and Flinn announced that the company would not grant unqualified recognition of the union, that the pressure could not be changed because twenty-one pounds was "the pressure allowed by law," and that the workers were demanding double wages, which was not true. The contractor also requested that police protection be provided at tunnel entrances. After about a week, the contractor agreed to increase pay to eight dollars and fifty cents a day for miners, with smaller increases for the other sandhogs. This was more than the strikers had asked for with regard to pay, but the pressure and working hours would remain the same.[9]

Work soon returned to normal, but a few days after the strike ended, on April 22, 1923, G. J. Slade, a Booth and Flinn lock tender in one of the New Jersey land shafts, died. The sandhogs continued to work, as they did following a small blowout in the New Jersey north tunnel going east on June 10. It was different, however, when another death occurred on August 20,

1923. Dennis Sullivan, twenty-eight, a subforeman for Booth and Flinn, was walking on the footway at the top of the New York river-shaft caisson when a handrail gave way. He toppled over the edge and fell more than sixty-five feet to the concrete floor, breaking every bone in his body. Sandhogs waiting their turn to enter the air chamber were unnerved and refused to work.[10] The work stoppage was only temporary, however, and work resumed without further incident through the end of the year.

Another important phase of the project began on August 20, as work crews for Rodgers and Hagerty, the winning bidder for Contract No. 5, began installing a power plant for excavation of the New York approach tunnels. The nearly $3.5 million, two-year contract also covered the construction of foundations for the New York ventilation building on the west side of Washington Street between Canal and Spring Streets. Demolition of buildings in the right of way began early in September and was completed by the end of October. By the end of the year, work on the Spring Street approach, the first to be excavated, was about 60 percent complete.

On December 1, 1923, the New York river caisson reached its final position, its cutting edge resting firmly on bedrock, but it would not be ready to receive the shields driven west from the New York land shafts until its bottom was sealed. The south New York tunnel was shut down on November 26, pending completion of the river shaft, and the north New York tunnel was shut down on December 21.

On New Year's Eve, the New Jersey north tunnel shield going east pierced the bulkhead of the river caisson, ready to start midriver tunneling. The New Jersey south tunnel going east had been driven only about halfway to the river shaft. By the month's end, shift supervisors reported 171 cases of caisson disease, with 14 of those cases suffered by engineering staff.

At the beginning of 1924, the only out-of-the-ordinary event down in the tunnels was a January 4 test in the New York south tube by radio experts from Westinghouse Electric and Manufacturing Company. Eighty feet below the surface of the river, the experts held up a loop antenna and turned dials on an experimental portable radio set (radios in those days were heavy, bulky, and anything but portable) until they heard music and bits of other programs from stations in Newark, New Jersey, and New York City. The men from Westinghouse were pleased, because they had proved that men working underground could receive emergency communications. New York Tunnel Commission Secretary Morris M. Frohlich, who thought up the experiment, happily announced after its completion, "The only thing now needed is a sender."[11]

Above ground, the year began with consideration of more fundamental issues. In January 1924, the tunnel commissions reported to their respective legislatures that the total tunnel cost would be about $14 million more than originally estimated, making the new estimate of "final" cost approximately $42,659,000. Part of the increase was attributed to 100 percent growth in traffic on both sides of the Hudson River, which required wider approach roadways and larger plazas, and to a six-inch increase in tunnel diameter to accommodate larger ventilation ducts. In addition, the cost of labor and materials had increased much more than anticipated. Holland also returned from his trip to Europe convinced that the tunnels needed expensive visual enhancements, such as interior tile lining and architectural treatments.

Most of the authorized funds had been spent, and now the commissions needed more. In an attempt to offset the bad news about increased costs by alluding to potential future benefits, the New Jersey commission's report stated, "This tunnel is without doubt the forerunner of a number of river crossings of this type, and the problems here met and solved, as in all engineering work, will reflect themselves in financial benefit to all successive projects."[12]

There was widespread agreement among engineers, urban planners, the public, and politicians that new interstate connections between New Jersey and New York were needed. There was no consensus, however, about which entity should build them. At the commissioners' direction, Holland had spent about six weeks between December 1923 and January 1924 studying the feasibility of a bridge or tunnel between Perth Amboy, New Jersey, and Tottenville, Staten Island. He recommended a bridge, to be completed in 1928. The commissioners were ready to take on the task, but they did not get the opportunity.

New York Governor Al Smith and New Jersey Governor George S. Silzer were fed up with the quarrels between the tunnel commissions that had slowed down the Hudson River tunnel, and they were concerned about the necessity of having a statewide referendum in New Jersey each year before that state's share of the construction funds could be made available. They appreciated the practicality of having future interstate links constructed and managed by a single agency that could finance the projects with its own bonds, secured by operating revenues without reliance on the credit of the states. In their annual messages to their legislatures in 1924, both governors urged that the Hudson River tunnel should be turned over to the Port Authority for completion and that the Port Authority should be entrusted with construction of all future interstate bridges and tunnels.[13]

The status quo had its supporters, however. The Motor Trucks Association of America favored letting the tunnel commissions complete the job, and the Republican legislators in each state were satisfied that the commissions were on the right track, particularly since Adams and other Democrats had been ousted from the New Jersey commission. In 1924, the legislatures decided, therefore, to let the commissions finish the Hudson River tunnel but to assign bridge-building responsibility to the Port Authority.[14]

The Port Authority immediately began plans to construct the Tottenville–Perth Amboy bridge (later named the Outerbridge Crossing) and another bridge between Elizabeth, New Jersey, and Howland Hook, Staten Island (the Goethals Bridge). J. A. L. Waddell, the engineer who had so angered Holland and the other tunnel engineers with his paper questioning their ventilation design in August 1920, designed both these bridges. The Port Authority was also authorized to construct a bridge across the Hudson River between Fort Lee, New Jersey, and 178th Street in Manhattan (later named the George Washington Bridge).

Even though the state legislatures assigned these bridge projects to the Port Authority, that did not mean that the tunnel commissions would be denied the right to construct other tunnels, or so they hoped. On March 1, 1924, their engineers were able to report that all seven ventilation shafts were sunk to grade, and three of the shields had passed through the river shafts on their way to a junction under the riverbed. The engineers expected the shield in the New Jersey south tunnel heading east to pass through the river shaft before the end of the month. The New Jersey north tunnel shield going west would not be started until after the New Jersey south tunnel going west had been completely driven, which was expected sometime in the middle of the summer. The engineers anticipated that, at the present rate of progress, the north tube's two segments would meet under the river by November 1. According to the *New York Times*, the engineers claimed that the problems they confronted at the beginning of tunneling had all been solved, and they were "now looking about for other similar tunneling jobs where the experience gained in this work will be of use."[15]

The engineers also claimed that growth in the science of deep tunneling was illustrated by the fact that there were no "serious" accidents from compressed-air operations and that the number of cases of the bends was so small as to be negligible. The work was still very dangerous, however, as they were soon reminded. On March 17, 1924, John Taggart of Booth and Flinn became the project's eighth fatality when a New Jersey land-shaft cage dropped on him unexpectedly. And it was only a few weeks until a serious blowout occurred. It started as a trickle of water, seeping through the roof just to the rear of the shield in the south tunnel, about sixteen hundred feet in from the terminus in Manhattan. At about 7:45 a.m. on Thursday, April 3, sandhogs supervised by foreman David Brown gathered around and discussed how best to stop the leak. In an instant, the trickle turned into a torrent, as water gushed into the air chamber from a twenty-foot-long, two-foot-wide gash in the wall. Brown yelled, "Run for your lives, men!" as his thirty-five-man crew dropped their tools and scrambled for the safety of the blowout curtain, two hundred feet away. The pressure in the tunnel was twenty-one pounds per square inch, and the men were already panting heavily when they passed the curtain at full speed, the water rising rapidly behind them. They did not stop to catch their breath until the tunnel began to slope upward, another one hundred feet or so further along. At that point, they were safe. On the river above them, a geyser shot up fifty feet as the air escaped to the surface. The column of water struck a small boat in the stern, spinning it like a top but not capsizing it. Fifteen Italian laborers on a wooden barge carrying cement were nearly tossed into the river as the geyser lifted one end of their craft in the air. They later said that for a second it seemed as though the timbers of the barge had been split apart.

An inspection revealed that the blowout had flooded about three hundred feet of the tunnel. Within a few hours, the air pressure was increased, and the water began to flow back out, until another blowout occurred at 11:45 a.m. This resulted in another geyser, alarming passengers on ferry-boats in the vicinity. The engineers made no further attempts to repressurize the tunnel until two barge-loads of clay were dumped over the spot from where the geysers sprang.

Holland told newspaper reporters that the blowout would put about two hundred men out of work for perhaps three days. But the damage was not serious, and such accidents were incidental to subaqueous tunneling. John McParland, head of the Compressed Air Workers of America, might not have agreed, had he been present, but he was in Albany trying to advance a bill to reduce the hours of work for sandhogs and to increase their pay. In his absence, about four hundred sandhogs on both sides of the river refused to begin their 4 p.m. shifts on April 9. They demanded a four-hour day instead of a six-hour day under the twenty-seven and twenty-eight pounds per square inch of pressure to which they were then subjected. Some of the men claimed that too many sandhogs were suffering from the bends and that there were twenty men then in the infirmary. As usual in these disputes, representatives for Booth and Flinn refuted the claim, saying that there were only "six or seven" in the hospital. The fact that they did not seem to know exactly how many were ill was reason enough for the sandhogs' concern.[16]

A few men reportedly went back to work the next day, perhaps because the strike was unauthorized with McParland out of town, but most stayed out. Representatives for the strikers asserted that forty injured men were recently carried out of the tunnels, and they would not report back to work in full force until safety conditions improved. Booth and Flinn's Chief Engineer M. E. Chamberlain stated that the work stoppage, which he refused to call a strike, was just an attempt to influence the legislators in Albany. "That there are thirty-five or more men in our private hospital is simply not true," he claimed. "We have no patients in our infirmary." Perhaps the "six or seven" patients that the company had admitted to the previous day had all been discharged overnight. "Working conditions," he said, "are ideal."[17]

As usual in such strikes, most of the sandhogs soon went back to work, and the digging resumed. It was only a month, however, before the next crisis occurred in a Booth and Flinn–built tunnel. This time it was in the commonwealth on the other side of New Jersey. In 1919, Booth and Flinn won the contract to construct two 5,889-foot-long vehicular and pedestrian tunnels through Mt. Washington in Pittsburgh. The first tube of the Liberty Tunnels was essentially completed in July 1922, but thanks in part to the studies that had been done by the Bureau of Mines for the Hudson River vehicular tunnel, it was recognized that the ventilation system would be inadequate. So the tube was closed to traffic while engineers worked to design a solution. By January 1924, both tubes were open and in use, even though the ventilation problem had not been solved. It was assumed that as long as the vehicles kept moving and there was adequate space between each vehicle, everything would be fine. Police officers monitored the traffic, and restrictions were in place to prevent the buildup of noxious gases until the redesigned ventilation system was in operation. The restrictions worked—for a while.

On Saturday morning, May 10, 1924, traffic was extremely heavy at the north tunnel exit due to a streetcar strike. It was not long before the traffic volume exceeded the tunnel's capacity, and all movement within the tube came to a stop. Signs posted in the tunnel instructed motorists to shut off their engines while standing still, but many drivers ignored the warning. People began grasping for air as the atmosphere grew foul and thick with fumes. A panic ensued, and many motorists abandoned their cars and began to stagger toward the exit. Some people passed out before they could make it to the end and had to be carried out by others. Only twelve victims were taken to the hospital, and no one died, but the north tube had to be shut down.[18]

A soon as Holland heard about the incident, he began calling engineers in Pittsburgh to find out what had happened. When questioned by a *New York Times* reporter, who wondered if an accident could cause a similar event in the Hudson River vehicular tunnel, Holland said, "It was not an accident that brought on the trouble, but the heavy congestion of unregulated traffic. In our tunnel, traffic will be regulated." He added that there were no police officers in the Liberty Tunnels, but there would be a police officer every five hundred feet in the Hudson River tunnel. "There will be no congestion of traffic here," he confidently stated, "because it won't be permitted to accumulate into congestion."[19] Holland then went over, in detail, the design of the Hudson River tunnel ventilation system, explaining to the reporter, as he had explained to so many other skeptics over the course of the previous few years, how well it had been thought out.

As chief engineer of the project, much of Holland's time was spent explaining the tunnel: explaining why certain design decisions were made, why the estimated final cost had increased, and why the completion date kept being extended. One of his most important job responsibilities was to serve as the tunnel's main booster and apologist when questions arose about problems with the project. In addition to answering questions from newspaper reporters, Holland conducted several radio interviews, published articles in magazines and trade publications, and delivered his lantern-slide presentation to groups such as the Brooklyn Institute of Arts and Sciences, the Detroit Engineering Society, and the General Science Alumni Association and student body at Cooper Union and to numerous citizens and professional groups throughout New York and New Jersey.[20]

Holland also gave tours of the shaft sites and, occasionally, of the air chamber to reporters, dignitaries, and other visitors, even though such public-relations efforts took up much of his limited time. The sandhogs did not particularly like having visitors come into their underworld realm, but they tolerated them. Once, Holland took his wife, Anna, down into the air chamber. The West Indians on that shift let it be known that women in the tunnel were bad luck, and she never returned.[21]

Anna Holland seldom knew when her husband would finish work for the day, and dinnertime was always uncertain. Almost every evening before Clifford Holland went home, he stopped by the tunnel site to check and see how things were going. On those rare occasions when Anna could induce him to take a break and go to the theater, he always went back to the tunnel afterward, spending hours in the field offices, personally supervising the work. He often went down into the air chamber on these visits, even though the process of going through compression and decompression was hard on his body. The beginning of compressed-air operations marked a general decline in his health, which grew more fragile as the two shields of the north tube inched closer and closer under the riverbed.[22]

Holland, perhaps fearful that a genetically weak heart might fail him before the job was complete, told his wife that one of his strongest wishes was to see the north tunnel "holed through." He knew that his mathematical computations and all the surveying and measuring had been correct, but there was always a possibility that when the time came, the two sections of tunnel would not line up. Even a small misalignment would be disastrous.[23]

Whatever damage had been done to Holland's body by the occasional physiological pressure of compressed air, the constant psychological pressure of his responsibilities had done even more harm to his mind. Sometime during the last week of September 1924, with the holing through of the first tunnel about a month away, Holland suffered a complete mental breakdown. He could no longer function and had to turn his duties over to Milton Freeman. On October 8, the tunnel commissioners quietly announced that the chief engineer would be taking a break. They granted Holland a month's vacation with full pay, and he was sent to John Harvey Kellogg's famous sanitarium in Battle Creek, Michigan, with the hope that he might recover in time for the holing-through ceremony planned for Wednesday, October 29.

The plan was to halt the movement of the shields when they came to within a few feet of each other. Dynamite would then be placed in the wall of earth separating the two tunnel segments. At the last minute, a telegraph wire would be attached to a detonator, ready to receive an electrical impulse transmitted from President Calvin Coolidge in the White House that would set off the explosives and blast away the barrier. Governor Smith and Mayor Hylan would be in the tunnel driven from New York, while Governor Silzer and Mayor Hague would be on hand in the tunnel from the New Jersey side, along with a reporter from Newark radio station WOR, which would broadcast the explosion as it occurred. Senators from both states would also be on hand to make speeches and to share in the glory of the event.

Last-minute calculations revealed that, as the two shields drew close to each other, the tunnels' alignment should be about perfect. As the *New York Times* reported, however, some of the engineers "lie awake nights just the same worrying about it." They need not have. When the holing through finally took place at noon on October 29, 1924, it was found that the relative positions of the tubes matched to within two one-hundredths of an inch.[24]

Holland would have been proud. When his former boss at the Rapid Transit Commission and good friend Robert Ridgeway visited Holland at the sanitarium in the third week of October, he found the chief engineer much improved, more enthusiastic than ever, and full of hopes and plans. "I shared his room that night and even after we had retired and the lights were out he kept on talking," Ridgeway later recalled. "Most of all he wanted to finish the tunnel. There was to be plenty of time for other things after that."[25]

Upon arriving back in New York, Ridgeway called Holland's wife and several of his friends to tell them how much better Holland was looking and feeling. But shortly thereafter, at 11 a.m., a telegram arrived. While undergoing a tonsillectomy on Monday, October 27, Holland died of heart failure on an operating table at the Battle Creek Sanitarium. He was just forty-one years old. His body was brought home to Brooklyn on the same day, at almost the same moment that the north tunnel was holed through. In his honor, the planned celebration was canceled. Booth and Flinn's President George H. Flinn, instead of President Coolidge, set off the explosion. After the smoke cleared, an electrician from Jersey City, A. F. Templin, was the first to crawl through the hole created, which was just large enough to admit his body.[26] After several other sandhogs passed through, a "photo opportunity" was set up as Superintendent Harry Redwood crawled up the slope of loose rock in front of him, reached through the hole, and shook hands with his brother, Norman Redwood, representing the New Jersey crew. The photographer took his shots, and after exchanging a few words, the men went back to their jobs. There was still work to be done.[27]

Top: curve in south tunnel; *bottom:* concrete construction in south tunnel

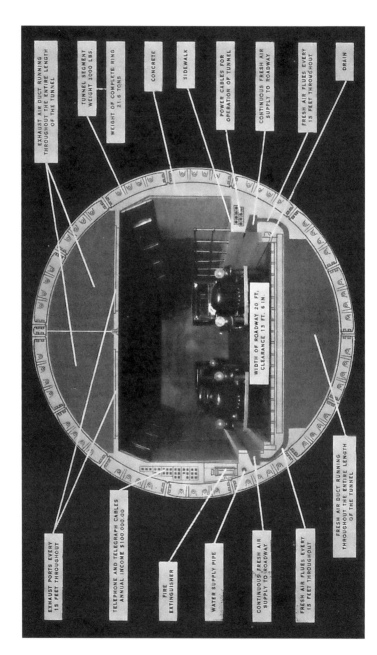

EXHAUST AIR DUCT RUNNING
THROUGHOUT THE ENTIRE LENGTH
OF THE TUNNEL

TUNNEL SEGMENT
WEIGHT 3000 LBS

WEIGHT OF COMPLETE RING
21.6 TONS

CONCRETE

SIDEWALK

POWER CABLES FOR
OPERATION OF TUNNEL

CONTINUOUS FRESH AIR
SUPPLY TO ROADWAY

FRESH AIR FLUES EVERY
15 FEET THROUGHOUT

DRAIN

WIDTH OF ROADWAY 20 FT,
CLEARANCE 13 FT. 6 IN.

EXHAUST PORTS EVERY
15 FEET THROUGHOUT

TELEPHONE AND TELEGRAPH CABLES
ANNUAL INCOME $100,000.00

FIRE
EXTINGUISHER

WATER SUPPLY PIPE

CONTINUOUS FRESH AIR
SUPPLY TO ROADWAY

FRESH AIR FLUES EVERY
15 FEET THROUGHOUT

FRESH AIR DUCT RUNNING
THROUGHOUT THE ENTIRE LENGTH
OF THE TUNNEL

Model of tunnel showing details

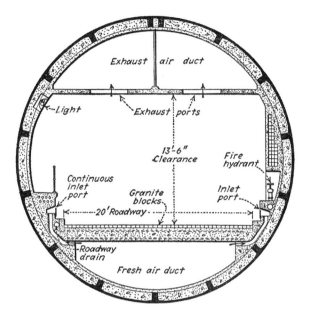

Top: cross-section of final tunnel design, 1927; *bottom:* special ventilation test apparatus erected at Hyde Park plant of B. F. Sturtevant Company

Right: state-line markers, north tunnel, 1926; *below:* Governor Moore, Mayor Hague, and New Jersey commissioners Kates, Noyes, Boettger, and Boyle outside tunnel, August 21, 1926

Emergency wrecking and fire truck, 1927

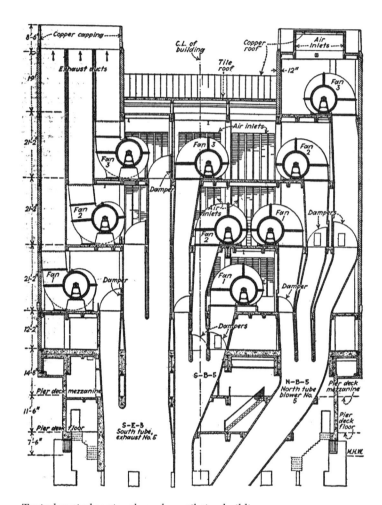

Typical vertical section through ventilation building

Top: New York river ventilation building, looking east, 1927; *bottom:* New York river ventilation building, looking west, 1987

Left: New York land ventilation building, 1987; *right:* New Jersey land ventilation building, 1987

Top: New Jersey river ventilation building, looking west, 1927; *bottom:* fresh-air duct under roadway, south tunnel, 1987

Above: New York land ventilation building, third floor, detail of blowers, 1987; *right:* blower fan, 1927

Top: blower air duct, New York land ventilation building; *bottom:* exhaust air duct, north tunnel

Top: opening ceremonies at Freeman Square, 1927; *bottom:* New York entrance plaza, November 14, 1927

Above: New York exit plaza, 1927; *left:* New Jersey entrance plaza, 1927

Top: New York entrance plaza, 1927; *bottom:* New York entrance, 1987

Top: burned truck in south tunnel, May 13, 1949; *bottom:* south tunnel, May 14, 1949

Top: north tunnel, looking east, ceiling removed, 1987; *bottom:* two-way electric-powered patrol cars tested in 1954

10

The Holland Tunnel

A tribute to the memory of the late Clifford M. Holland, the first chief engineer of the work, is made part of the [1925 New York Tunnel Commission] report, along with the statement that the tunnels had been named "the Holland Tunnel" in his honor.

—*New York Times*, March 22, 1925

THE REQUESTS WERE not long in coming. Many professional societies, civic organizations, and prominent citizens expressed to the tunnel commissions their desire that some special honor be given Clifford Holland in recognition of his dedication and sacrifice. The New York Section of the American Society of Civil Engineers, of which Holland had been a founding member in 1920, specifically asked that the Hudson River vehicular tunnels be formally renamed "The Holland Tunnels" in honor of the deceased chief engineer. At meetings held on Tuesday, November 12, 1924, both commissions resolved that the new designation "The Holland Tunnel" (singular, not plural) be adopted for the facility. This honor was not revealed to the public, however, until the following March, when the commissions released their annual reports for 1925. In the meantime, the project continued to be known as the Hudson River Vehicular Tunnel.

Milton Freeman, who had been acting chief engineer since Holland's breakdown, now directed the project. He was the perfect man for the job. As engineer of construction since July 1, 1921, he had the responsibility to see that the building of the project was executed according to plan and specifications. He was officially appointed chief engineer on December 1, 1924. Under his supervision, the last shove of a shield in the north tunnel took place on December 9, 1924, compressed-air operations halted on January 3, and the last regular cast-iron ring erected on January 27, 1925. Installation of a special junction ring to span the gap between the New York and

New Jersey segments on June 30 marked completion of the outer lining of the north tunnel.

The south tunnel headings were holed through on December 7, 1924, with the last shove taking place on January 10, compressed-air operations stopped on February 22, and the last regular ring erected February 28, 1925. Because there was less of a gap between the headings in the south tunnel, the space was closed with special wedge-shaped steel plates on March 15.

The last shove of the New Jersey north tunnel going west toward the open-cut approach took place on March 6, 1925, and compressed-air operations were discontinued March 23, marking the completion of major tunneling. This last phase of work had been slowed considerably by the need to slowly and carefully push underneath an old sewer in the Erie Railroad yard at the end of Twelfth Street in Jersey City.

The work of lining the tunnels with concrete began in 1924 between the land and river shafts in New York and continued through 1925 into 1926. As soon as the air pressure could be lowered to normal levels in the river segments of each tunnel, workers demolished the concrete bulkheads forming the back of the air chamber and dismantled the air locks and other equipment. The contractor's plants and equipment in Canal Street, in Spring Street, on the land shafts, and on Pier 35 in New York were removed, and the powerhouse was dismantled and the site razed. Almost all the 174,626 cubic yards of protective clay blanket over the tunnel were removed from the riverbed, and the lease of Pier 35 was terminated. A new lease for a twenty-foot-wide access roadway over Pier 35 to the river shaft was arranged, as were easements across the yard of the Lackawanna Railroad in Jersey City.

Freeman had been working hard before Holland's breakdown, and now, with the full weight of responsibility on his shoulders as chief engineer, he worked even harder. Like his predecessor, he put in long hours at the office and at job sites, frequently skipping meals and napping on a couch in his office. On March 17, just two days following installation of the last plates of the south tunnel, he became too ill to work and had to go home. He never returned. At about 10 p.m. on Tuesday, March 24, 1925, he died of acute pneumonia at his residence in Valhalla, New York. He was fifty-five.

There were four other deaths that year. In March, Feodor Tarashiep and Sezoy Palischick were killed during the razing of a building on Canal Street. August Nevoia, an employee of Booth and Flinn, died after being struck by the cage in the New Jersey river shaft on June 8.[1] A fourth man, whose name is unknown, died of bronchial pneumonia while in the hospital under treatment for partial paralysis brought on by caisson disease. Because it was assumed that the victim contracted pneumonia in the hospital and not in the tunnels, and because caisson disease was listed merely as a secondary cause of death, this man was never listed among those who died working on the project. He should be. Even if his pneumonia was not brought on by work in compressed air, he would not have been in the hospital had he not suffered from caisson disease. If it can be said, as it was, that Holland and Freeman sacrificed their lives for the project, surely the same can be said of this sandhog.[2]

On April 7, 1925, the commissioners appointed Ole Singstad chief engineer, and he took over all the burdens that had contributed to the early demise of Holland and Freeman. He was on the job just two weeks when approximately 180 sandhogs went on strike at noon on April 21, 1925. They had been restive since the cessation of compressed-air operations in the main tunnel segments, when their pay had been cut by about 50 percent to match that of ordinary laborers. Booth and Flinn requested that police officers be stationed at both tunnel entrances and exits. Because the sandhogs were no longer needed and their places could be filled easily with new men, the strike accomplished nothing.

The tunnel commissions awarded eight new contracts in 1925, covering tile and tunnel finish (no. 7); furnishing power transformers and oil switches (no. 9); tunnel roadway paving (no. 10); fans, motors, transmissions, and control systems (no. 12); ventilation buildings and interior shaft construction in New York (no. 13); ventilation buildings and interior shaft construction in New Jersey (no. 14); tunnel power cables (no. 20); and pumping equipment (no. 24).

A few other contracts were ready or almost ready by the end of 1925 but were not awarded until 1926. These included contracts for architectural work at the New York entrance and exit plazas (no. 15); architectural work at the New Jersey entrance and exit plazas (no. 16); the New York emergency-equipment building (no. 17); the New Jersey emergency-equipment building (no. 18); lights, traffic signals, supervising system, and tunnel and plaza wiring (no. 19); emergency trucks and equipment (no. 22); plaza pavement in New York (no. 23); fire extinguishers (no. 25); and plaza pavement in New Jersey (no. 26).

Three of the contracts awarded in 1925 covered construction of the all-important ventilation system, the operation of which would mean success or failure for the tunnel. The land ventilation building in New York would be five stories and 122 feet high, while the New Jersey land ventilation building would sit on a raised deck and be 115 feet high. Both river ventilation buildings would rise 107 feet above the pier decks. The drawings for all these buildings were submitted to the Fine Arts Commission of the State of New York and revised according to suggestions made by the commission. As a result, these structures would have some ornamental details that somewhat mitigated their otherwise utilitarian appearance.

There would be a total of fourteen sets of intake and fourteen sets of exhaust fans in the system, each fan measuring eight feet in diameter. The intake fans would draw fresh air in from outside through louvers located along the ventilation building's exterior walls. The air would then be divided among fourteen fresh-air ducts and blown into the seventeen-foot-wide main duct underneath the roadway in each tunnel. From these ducts the fresh air would pass through ports located ten to fifteen feet apart along the curb of each roadway, where it would mix with vehicle exhaust before being drawn up toward the drop-down ceiling. Adjustable slides over the openings would control air flow over the curb ports.

The exhaust fans would draw out the foul air through slits in the ceiling into the twenty-one-foot-wide, domed exhaust air duct. Fourteen exhaust air ducts would channel this air into the ventilation buildings, where it would be expelled to the outside. Each duct would be equipped with three adjustable-speed fans, two of which, when operated together, could supply the needed quantity of air, with the third serving as a backup. The system had the capacity of handling fourteen hundred tons of air per minute, with a complete change of air in each tunnel forty-two times per hour.

The fans, of which there would be eighty-four in all, were B. F. Sturtevant Silentvane centrifugal units. An early version of this type of fan was developed in 1922 and soon proved to be much more efficient than anything similar on the market. It became very popular among architects, engineers, and industrial users. The Sturtevant company engineers and technicians worked closely with the tunnel engineering staff to select the proper type of fan and to determine the frictional resistances against which they would operate so that motors of proper capacity could be installed.

In addition to the input sought from the state's Fine Arts Commission regarding aesthetic treatment of the building exteriors, a color psychologist was engaged to consider the possible use of tints in the otherwise white interior tiles. The advice of this expert was that all traces of blue, green, or red should be eliminated because of their depressing effects, although a light orange was selected for the borders. American Encaustic Tiling Company of Zanesville, Ohio, furnished the tile for the north tunnel, while the tile for the south tunnel came from companies in Germany and Czechoslovakia. The sixteen-month contract for tile work (no. 7) was awarded May 6, 1925, but due to quality-control problems, this work took much longer than expected.

At the beginning of 1926, it was hoped that all construction and equipment installation would be nearly complete by the end of the year. With a reasonable amount of time allowed for testing of equipment and operating procedures, the tunnel should be ready for public use early in 1927, perhaps as early as February. The only remaining holdup was the need for additional funds.

The tunnel commissions, in their reports to their respective legislatures in January 1926, asked for another $3,283,300 to complete the project. As unwelcome as this latest increase in cost may have been, automobile-traffic counts provided by the ferry companies indicated an overall increase of about 10 percent per year, and it was believed that the tunnel would be able to capture more than enough of that traffic to reimburse its cost ahead of schedule.

How the street systems of Jersey City and New York were going to handle all the traffic going through the tunnel was of increasing concern as the time for completion drew nearer. The Board of Estimate of New York City voted on June 14, 1926, to approve general plans for construction of an elevated "West Side" express highway extending from the exit plaza of the tunnel at Canal Street to Riverside Drive at Seventy-Second Street, a distance of about four miles. The project, expected to cost about $13.5 million, was an initiative of Manhattan Borough President Julius Miller. As planned, the highway would carry six lanes of traffic, with ramps about every ten blocks for access to cross streets at grade level. The plan had support from the Port Authority, the Sinking Fund Commission, the Fifth Avenue Association, and various other civic organizations. It also had its detractors. The Municipal Art Society of New York stated that the highway would be "unsightly and calculated to interfere with surface traffic." Mayor James J. "Jimmy" Walker, who took office in January 1926, also opposed it because it might block access to the waterfront. He also felt that once the tracks of the New York Central Railroad were removed from Eleventh Avenue, the highway might not be necessary.[3] Opposition from Walker and others resulted in project delays, and the West Side Elevated Highway was not completed until well after the tunnel began pouring traffic onto city streets.

In the short term, New York City devised a plan to extend Sixth Avenue from Carmine Street to Laight Street to facilitate traffic movement to the north. The new street would be about thirty-two hundred feet long and one hundred feet wide, and it was estimated to cost the City about $500,000 for construction and $3.62 million for acquisition of property. It eventually cost more than $4.5 million for land alone.[4] In addition to the cost, the extension displaced about ten thousand residents and shopkeepers, most of them lower-income Italians. The City of New York Finance Department had notified these people that they needed to find new places to live and new sites for their businesses, but most of them did not believe it. "This talk of razing the houses is not new," said one man who had lived in the neighborhood for twenty-five years. "It has been going on for nineteen years and nothing has ever happened. Consequently, when we were told that the City was going to tear down the buildings many of us regarded it as just more talk."[5] On August 31, 1926, the last day of the grace period expired, and the City informed these citizens that they would be subject to eviction.

Among those who might suffer due to their delaying departure were the parishioners of the Catholic Church of Our Lady of Pompeii, at 210 Bleecker Street. They were constructing a building at a new location, but it would not be ready in time. Close by, occupying one corner of Bleecker and Carmine Streets, the firm of Petrick, Schmitt and Bergman, printers and lithographers, had machines weighing from five to twenty tons that could not be relocated easily. Julius Moscovitz, a clothier located on another corner of Carmine and Bleecker Streets who lived above his shop, was building a new home a few blocks away, but it would not be ready for another sixty days. "These are poor people down here and they are not finding it easy to get new apartments with cheap rents," Moscovitz said. "As for the storekeepers, many of them would literally be on the streets if an eviction order should be issued," he added.[6]

Some of the other business owners had waited until the last minute to close down operations, but they had met the deadline. Morgan's Oyster House on Spring Street was a landmark restaurant frequented by politicians and actors. It was to have celebrated its fiftieth anniversary on August 31. Instead, motion-picture photographers filmed its last day of business before it closed on Monday, August 30.

Nearby, at the tunnel entrance, the engineers had tried to limit property destruction to the least valuable tracts. But their decision would diminish the already fading domestic charm of the neighborhood. The block bordered by Hudson Street on the west, Varick Street on the east, Dominick

Street on the north, and Broome Street on the south would be split by the open-cut entrance to the westbound tube. Here low-rise residences with old-fashioned stoop entrances, doorways marked by fluted Ionic columns, and big dormer windows protruding from the attics served as examples of the homes once common to the area. As one older resident of Dominick Street recalled, speaking of how the neighborhood used to be, "As nice a lot of people as any one could wish to meet lived in these streets then, and they have not all gone yet, but the new tunnel will probably drive us out."[7]

The entire block to the south between Hudson, Varick, Broome, and Watts Streets would eventually be cleared of all structures, but many of these were dilapidated two-story buildings filled with junk shops on the lower floors. A few old stables also remained. As the *New York Times* observed, the western half of this block "presents a pitiable object lesson of the ruins of lower Greenwich Village."[8] Not everyone found this block of little value, however, and after a number of Watts Street property owners protested plans to take at least a portion of their property, Borough President Miller sent a letter to the newly appointed Board of Estimate in February 1926 asking that the widening be delayed. He later stated that he did so to lessen opposition to the Sixth Avenue extension plan, with hopes that traffic congestion at the tunnel entrance would not be all that bad. William Bullock, director of the City Affairs Bureau of the New York County Republican Committee, thought there was another reason. He charged that the delay, agreed to by the Walker-appointed Board of Estimate, was done at the request of James J. Riordan, treasurer of the campaign committees for Walker and Governor Smith in 1926 and a close friend of both men.[9]

Riordan was founder and first president of the United States Trucking Corporation, which paid Smith a very generous salary to serve as its chairman of the board in 1921 and 1922, when he was between terms as governor. The corporation owned, or controlled through a long-term lease, property on Watts Street that would have been taken had the street-widening plan not been canceled, thus leaving the company without a freight warehouse and transfer facility virtually at the mouth of the tunnel. Miller claimed to be ignorant of this, and friends of Riordan claimed that he no longer had an interest in the company and had no interest when the widening plan was rescinded. Maybe that was true. Riordan had resigned as president of the trucking firm to organize the County Trust Company of New York, with Smith as director and his first depositor. He opened the new company on February 3, 1926, just days before the new Board of Estimate voted to cancel the street-widening plan. Of course, just because he was no longer an officer did not mean that he had no financial "interest" in the trucking firm. He also owned a large amount of real estate in Greenwich Village.[10]

In Jersey City, there were fewer privately owned properties subject to adverse effect by tunnel traffic, but one important public property was imperiled. In September 1926, a fight developed between the New Jersey State Highway Commission and the Hudson County Park Board over a plan for the main road leading to the tunnel. This highway would connect the tunnel with the prospective bridge from Fort Lee to midtown Manhattan and to the Staten Island bridges. The Highway Commission wanted to run the road through West Side Park (now Lincoln Park), the largest of seven parks supervised by the Park Commission. The park commissioners strongly objected because to do so would ruin pending plans for a golf course. At the suggestion of Hague, a depressed roadway section would be considered as a compromise.[11]

New Jersey was also making plans for a thirteen-mile-long, $40 million, four-lane express highway connecting the tunnel plazas with the Lincoln Highway. Segments of that highway were already under construction by the end of 1926, but disagreements about how to span the Hackensack and Passaic Rivers stalled further progress.[12]

Unresolved challenges regarding potential traffic did nothing to impede progress on the Holland Tunnel, which was complete enough by August 21, 1926, for New York Governor Smith and New Jersey Governor Moore to take an official tour of inspection. At 11:45 a.m. that Saturday morning, two buses, several private automobiles, a truck carrying reporters, photographers, and motion-picture camera operators, and a police motorcycle escort entered the westbound tube's New York entrance. A crowd that had gathered around the entrance sent up a cheer as the vehicles disappeared into the tunnel.

The two governors, the members and staff of each tunnel commission, and the principal staff engineers rode in the first bus. The second bus carried another twenty-eight or so people, including Mayor Hague, Senator Edwards, James Riordan, George Flinn, Edward Byrne, and several relatives of New York Commission Chairman Dyer. The private cars carried the governors' family and friends. It took but a short time to reach the exit in New Jersey, even though the vehicles had to proceed slowly at two points where wood planks covered unpaved portions of the roadway over drainage sumps. As the procession emerged into the sunlight in Jersey City, people lining both sides of Fourteenth Street, including five thousand children clutching American flags, cheered and waved as the vehicles passed by. After looping through several blocks of the Horseshoe neighborhood, the line of vehicles entered the eastbound tunnel and headed back toward New York. Reaching the blue-tile line that marked the point of division between the two states, the vehicles stopped as everyone piled out to watch the ceremony. The governors and commissioners stood underneath the embedded signs in the tunnel wall demarking their states, while the camera operators and photographers set up their equipment. Then Smith reached out his hand, grasped Moore's, and with a smile and a gentle yank said, "Come on over into New York State." They shook hands for a while in the harsh glare of a light mounted on the press truck as the news reporters shouted, "Hold it!" and took their photographs. Then Moore pulled on Smith's hand as he said, "Come on over to New Jersey. I will show you a good state." After more smiling and hand shaking, everyone reboarded their vehicles and drove back to Manhattan.[13]

On the day of the inspection tour, the commissioners and engineers projected that the tunnels would be open to traffic by March 1927. They were overly optimistic. The project was far from complete, and there was still much work to be done. Some of it was very dangerous, and before the year was out, two more workers would die. On October 28, Charles Svenson, an employee of DeRiso Construction Company, died immediately after being struck by an eight-foot maul in the New Jersey river ventilation building. Fewer than two months later, a contractor became the final project fatality. Shortly after 4 p.m. on Monday, December 20, James G. Godfrey, fifty, of the Albee-Godfrey Whale Creek Construction Company, led one of his foremen, George Ross, and a young apprentice into the shell of the New York south river ventilation building. The three started a ninety-five-foot descent to the bottom of the shaft down a spiral wood staircase, Godfrey lighting the way with a small flashlight. Fifteen feet from the top of the staircase there was a landing, onto which Godfrey stepped while his companions were still on the staircase. Suddenly the light went out, leaving Ross and the boy in total darkness. From below came a single cry and a splash as Godfrey hit a pool of water at the bottom of the shaft. Not daring to move, Ross called for help until someone finally heard him and turned on some lights. Before him, Ross could see the broken plank through which Godfrey had plunged. A fire department emergency truck and an ambulance were called, but only the latter was needed to take the recovered body to St. Vincent's Hospital.[14]

The cost of tunnel construction in terms of human life did not rise, but the cost in dollars did go up once again. On December 26, 1926, Dyer announced that the total cost of the project would now be $48,400,000, or $5,741,000 more than the revised estimate made in January 1924. Each state would be called on to make a final appropriation of $1,650,000, but Dyer said that the states would have a new source of income of about $9 million a year once the tunnels opened. He also projected that the total project cost could be recovered in ten years instead of twenty and that after the costs were recovered, the tolls would be abolished.[15]

The state legislatures, as they had done with previous cost estimates, accepted the revised figures and appropriated the necessary funds. The expenditures would be justified, they assumed, by a copious flow of revenue into state coffers. That revenue would come from collections at eight toll booths in Manhattan and six in Jersey City. Each booth would be made of steel set in copper sheathing and placed between lines of vehicular traffic. The construction contract for these booths was one of the project's last.

The projected revenue stream would flow only if the tunnels worked as planned. But with the projected date of opening just months away, the ventilation system was not yet operational, and there were many people who still doubted that it would work. It was not until Tuesday, March 15, 1927, that the engineers were finally ready to test the system. With senior staff and Consulting Physician Edward Levy in tow, Singstad led members of both commissions on a tour of the New York land ventilation building. Accompanying the group were three special guests: William T. Donelly, W. J. L. Branham, and John F. O'Rourke of the New York Board of Trade and Transportation. After examining one of the airtight exhaust chambers and the fans within it, the men went down into the south tunnel and walked through one section. They inspected the fire and flushing hydrants placed at 240-foot intervals and the chemical firefighting apparatus. With two exhaust and two intake fans running and a 650-foot section of tunnel blocked off by dampers, the engineers set off two three-minute smoke bombs, which Singstad claimed would emit a volume of gas and smoke equivalent to that from hundreds of vehicles or "a good-sized fire." Some of the thick yellow smoke drifted into adjoining sections of the tunnel, but most of it disappeared into the exhaust slits overhead within a few minutes. Two more bombs were then set off in the air chambers behind the curbs of the roadway, with similar results.[16]

The test, according to Singstad, was a success. The representatives of the Board of Trade and Transportation did not agree. On April 13, the board adopted a resolution calling on the tunnel commissioners to conduct more stringent tests. This action was a response to a report written by a special investigating committee headed by O'Rourke, who thought that the tests conducted by Singstad did not approximate conditions that would follow a vehicular crash. "In our opinion," the report stated, "a satisfactory test of an ordinary fire should involve the burning of at least fifteen gallons or approximately 100 pounds of gasoline."[17]

The report may be seen as evidence of O'Rourke's continuing antipathy to the project due to his exclusion from any of the construction contracts, but he had a good point. The test proved that the system, or at least a portion of it, could reasonably handle the smoke from a few bombs when the tunnel was free of traffic. In order to calm widely held fears that the system was inadequate, a true test was needed under conditions more closely approximating those that might be encountered when the tunnels were in operation. Such a test should include an evaluation of firefighting equipment and emergency procedures. As Board of Trade General Manger Charles J. Columbus wrote to Singstad on April 21, "The fears expressed in the board's report are felt by people generally and are supported by the opinion of experts in ventilation, engineering and fire fighting." Before that type of test could take place, however, a police force would have to be on hand and equipment available.[18]

On May 15, former New York City Deputy Chief Police Inspector Cornelius F. Cahalane announced that he had been hired by the tunnel commissions (as a consultant) to organize a police force and that half of the officers would be selected from New Jersey applicants and the other half from New York applicants. Special legislation would have to be enacted in each state to address the jurisdictional divide that existed in the middle of the river, but until then, New Jersey laws would be enforced on the New Jersey side and New York laws would be enforced on the New York side. Plans were also announced for equipping the tunnels with four electric emergency wagons, one stationed at each entrance or exit.[19]

With a police force hired and trained and proper emergency equipment on hand, the engineers conducted a final test of the ventilation system on Thursday, November 3, 1927. Accompanied by representatives of the New York City and Jersey City fire departments and news reporters, the tunnel commissioners and engineers walked into the north tunnel from the New Jersey side and down about one thousand feet, where a decrepit touring car awaited its fate. After Singstad explained what was about to happen, the observers were asked to stand back while six gallons of gasoline were poured on the roadbed underneath and around the car and another gallon poured onto the upholstery. One of the engineers then tossed a match into the vehicle, which instantly erupted in flames that spread to the fuel on the road. A patrol officer stationed nearby pushed an emergency button, which set off an alarm in the New Jersey exit station and automatically activated signal lights which would warn drivers in a real emergency to move over into the right-hand lane so that emergency vehicles could use the left-hand lane. The officer grabbed a five-gallon fire extinguisher from its nearby cache as two more officers, one from the station to the east of the fire and one from the station to the west, raced to the scene. One officer handed extinguishers from the pedestrian walkway down to his fellow officers below, as the flames rose to the ceiling and thick black smoke billowed up from the sacrificial automobile. It took four extinguishers and about three and a half minutes to put the fire out. The ventilation system seemed to work well enough, and most of the smoke was sucked away quickly through the overhead vents.

Before the fire was fully extinguished, one of the double-end-drive emergency vehicles roared up. To demonstrate the firefighting ability of this combination fire truck, wrecker, first-aid station, and ambulance, more gasoline was poured onto the car, and it was set alight once again. This time it took the three-man emergency vehicle crew just two minutes and fifteen seconds to put the fire out. Within another thirty seconds, they had the car hooked up and began to tow it out of the tunnel.

After the demonstration, Dyer told the news reporters that if every person in New York and New Jersey had been able to witness the efficiency with which the fire was extinguished, then "tunnel fear" would vanish as quickly as had the flames. But when a reporter asked one of the New York commissioners about another burned and battered wreck standing outside the entrance to the tunnel, he was told, "Oh that? We burned that one in the tunnel this morning just to make sure that the demonstration would go through all right this afternoon." Apparently, the commissioners had had their own doubts about whether the system would work properly.[20]

The last remaining issues to be addressed before the Holland Tunnel could be opened to traffic concerned the type of vehicles that would be allowed to use the facility and the toll rates to be charged. When enabling legislation was first enacted in 1919, it was assumed that the tunnel would be used by pedestrians, horse-drawn wagons, "motor-trucks," and a few "pleasure vehicles," meaning passenger cars. For several years, it had been apparent that there was no longer sufficient demand for use of the tunnel by horse-drawn wagons, as this mode of transportation was rapidly fading away. Besides, horses were slow and prone to cause trouble, so horses were banned. It was also recognized, partly due to what had happened to pedestrians in the Liberty Tubes, that there were significant safety concerns with allowing large numbers of people to slowly walk trough the tunnel from one state to the other. Before that event, it was thought that the toll rate for pedestrians would be set high enough to discourage this type of traffic. Now it was decided to prohibit pedestrians. Bicycles would also be prohibited because of their perceived incompatibility with motor-vehicular traffic.

The toll rate for buses turned out to be more of an issue than anyone had anticipated, as this mode of transportation rapidly grew in importance. Hague joined bus operators in requesting a low rate, saying that a low rate would encourage growth in Jersey City. The request was offset by those who wanted to protect passenger traffic on the ferries and by Joseph E. Keen of the New York Central Mercantile Association. "The tunnel was built for the shippers and receivers of merchandise," Keen reminded the commissioners. "There should not be a favored rate for buses."[21] Keen was right about one thing. The tunnel had been built, according to its original justification, primarily for the movement of freight. But the world had changed since the tunnel was conceived, and there was now no doubt that privately owned automobiles and commercially operated buses would be using the tunnel in great numbers.

The commissioners wanted to set rates for all vehicles at a level low enough to encourage use of the facility and to generate enough revenue to reimburse the states within ten years. But they did not want to establish rates that were so low they would drive the ferries out of business or encourage the railroads to completely abandon the car-float and lighterage system. That would only substitute one mode of transit for another. Moreover, it would overburden the tunnel and lead to congestion during normal operation and deprive the public and shippers of alternatives should the tunnel ever be shut down due to an emergency. There were rumors, however, that the railroads were already making plans to ship all their freight through the tunnel.

On November 1, 1926, more than a year before the tunnel opened for traffic, Wilgus invited New York Commission Counsel Windels to lunch to call his attention to the threatened use of the tunnels by railroad corporations for freight trucking, to the possible exclusion of other vehicles. In September 1927, the railroads informed the tunnel commissioners that they expected to push the tunnel beyond design capacity on the day it opened. The Erie Railroad had already completed construction of a $5 million cold-storage plant and transfer facility alongside the new concrete and steel viaduct built to carry tunnel traffic westward from Jersey Avenue to Bergen Hill.

On September 16, 1927, Singstad informed the members of the traffic committee of the Broadway Association, "All the railroads which at present have terminals on the Jersey side and bring freight to Manhattan on car floats and lighters will attempt to save the expense of lighterage by telling shippers and consignees that that they will have to send motor trucks through the tunnel to the Jersey City freight yards to ship and receive goods." It was also anticipated that there would be heavy use of the tunnel for transportation of fruits and vegetables into Manhattan by non-railroad-related shippers.[22]

In response to this threat, the commissioners announced toll rates on November 1 that would be less expensive than the ferry rates for private automobiles of less than seven-person capacity and less expensive for buses but more expensive for trucks. Furthermore, trucks exceeding fifteen tons in gross weight would have to apply for special permits more than twenty-four hours in advance.[23]

One bus company operating out of Jersey City and holding a franchise to use the tunnel claimed that the rates for buses, though relatively cheap, were too high. But J. A. Hoffman, vice president of the Motor Haulage Company of Brooklyn, which claimed to be the second-largest trucking business in the city, claimed that the commissioners were trying to restrict the tunnels to use by "pleasure cars" and buses. It was his opinion that automobiles had little or no business so far downtown as the Manhattan entrance to the tunnel, and he had always understood that the tunnels were to be mainly for use by commercial vehicles. Other representatives of the trucking industry made the same charge.[24]

The commissioners denied that they had any desire to discourage truck traffic and stated that the tunnel was built to accommodate both trucks and passenger cars. In a joint statement released November 4, the commissioners cautioned, "What the public should bear in mind now is that the Holland Tunnel is a new proposition. There is nothing like it in the world. It is only by its practical operation that we can discover whether the rates charged are the fairest that can be fixed or whether there should be changes in them." In other words, once the facility was open and in use, adjustments could be made, but the rates and regulations would stay the same for the moment.[25]

There were, in fact, many questions regarding tunnel operation that would not be answered until the facility was in use, data collected, and problems identified. After all the planning and preparation, no one knew for certain if the traffic and revenue projections would be accurate, the approaches and roadway linkages sufficient, the basic design and equipment correct, and the personnel properly trained. But the day of truth was just ahead. On November 12, 1927, eight years after passage of the enabling legislation and seven years after construction began, the world's longest underwater tunnel would finally open.

11

One Work Complete

He who has made on earth one work complete
To live beyond this little span of days
That we call life — at parting of the ways
Has left a monument before our feet
At which we bow in reverence. Thus we stand
So near in homage we can clasp his hand.

—from a memorial tribute to Clifford Holland, by
Marguerite Janvrin Adams, the wife of one of
Holland's Harvard University classmates

"I COULDN'T WAIT to enjoy my tunnel, and so when the speeches were over, I set out to walk through it alone. Soon I heard a rumbling, shuffling sound in the distance. 'Good God!' I thought, 'it sounds like an ocean, like the tunnel's caved in!' I jumped up to the sidewalk." From this vantage point, Chief Engineer Ole Singstad was relieved to see that the source of the noise was not a wave of water but a wave of pedestrians.[1]

Made nervous by predictions of disaster, the tunnel commissioners had decided to delay opening the Holland Tunnel for vehicular traffic until 12:01 a.m. on Sunday, November 13, 1927. Pedestrians, however, would be allowed to walk through the tunnels from 5:00 to 7:00 p.m. on Saturday, November 12. At 4:55 p.m. in the early evening, with a touch of the gold and platinum telegraph key used by President Woodrow Wilson to explode the charge that opened the Panama Canal, President Calvin Coolidge sent an electrical current from his yacht, *Mayflower*, at anchor in the Potomac River. After traveling from Washington to New York, the current rang bells on the speakers' stands located at the tunnel entrances. The bells were supposed to signal workers to draw aside huge American flags from each portal, but few were left in Jersey City to hear the ring. The crowd on that side of the river had already entered the eastbound tube—about an hour and a half early.[2]

For the sake of politics, all the speeches given at the entrance to the eastbound tunnel in Jersey City by Governors Moore and Smith, Mayor Hague, Commissioner Boettger, and Senator Edge had to be repeated at the entrance to the westbound tunnel in Manhattan. That was just as well; little of what was said could be heard in Jersey City due to the incessant shrieking of locomotive whistles from the nearby Erie Railroad yard. Soon after the automobiles and buses carrying politicians, dignitaries, and newspaper reporters disappeared into the south tube heading east to Manhattan, the police decided to let the restive crowd enter the tunnel rather than keep them waiting—thus the early start of the pedestrian parade from that side of the river.

When radio reports announcing the end of speechmaking in Jersey City reached the speakers' stand in Manhattan, Commissioner Dyer was heard to say, "Thank God that's over."[3] It was not long before the party from New Jersey arrived at the New York plaza, which had officially been named "Freeman Square," to join the New Yorkers on the speaker's platform. Following the invocation by Bishop of New York William T. Manning, Governor Moore repeated the brief speech he delivered in Jersey City, hailing the magnificence of the engineering achievement in light of "the grave difficulties encountered in the course of construction." Governor Smith spoke next, stressing that the tunnel would forever eliminate danger from food and fuel shortages. Singstad, who did not speak at the Jersey City celebration, took to the stand at Commissioner Dyer's invitation and emphasized that research done during the development of the ventilation system would serve as the basis for economical design of future vehicular tunnels. He then slipped away to enjoy what he hoped would be a few moments of peace in "his tunnel" prior to its opening.

Senator Edge, in reference to Smith's characterization of the ceremony as the "wedding of two Commonwealths," said, "May that wedlock always remain and may it never be severed by a divorce." Goethals, who had only a few weeks left to live, did not attend the ceremony. Given the bitterness he bore to the end of his life over not being allowed to build the tunnel, he would not have appreciated hearing Edge conclude the speeches by praising the Holland Tunnel as "second only to the Panama Canal, if second at all."[4]

By 5 p.m., with the ceremonies over, the bells rung, and the flags pulled aside, both tubes of the Holland Tunnel officially opened for inspection by the citizens whose tolls would pay for the project. As Singstad stood on the raised walkway and looked down, he witnessed thousands of people, some pushing baby carriages, pass by on their way from one state to the other. A holiday spirit prevailed as the crowd laughed and sang while they walked. Some shouted just to hear the echoes of their voices reverberating off the tiled interior walls. Patrol officers laughed and joked along with the strollers, who filled the roadway from curb to curb. Many people stopped and bent over, hands outstretched, to feel the air as it gently whooshed through the vents at their feet. Upon reaching the blue-tiled demarcation line identifying the boundary between the states, they shook hands with each other, some standing in New Jersey while grasping the hands of others standing in New York. They also discovered two points, one under each river ventilation building, where persons in one tube could shout greetings through the connecting emergency passageways to persons in the other tube.

New Jersey tunnel commission secretary E. Morgan Barradale later estimated that approximately fifty thousand people from New Jersey walked through the tunnels, and another twenty to twenty-five thousand from New York passed through before the tunnels were closed to pedestrians at 7 p.m. The difference in numbers may have reflected the relative importance placed on project completion by residents of the two states. In Manhattan, observation of the opening was restricted mainly to events on Saturday, which included a private luncheon at the 71st Regiment National Guard Armory at Fifth Avenue and Thirty-Fourth Street, a parade to the tunnel, and the ceremony at the entrance plaza. Across the river in Hudson County, the widespread festivities went on for days.

Flags flew from many buildings, and businesses and private homes throughout Jersey City displayed red, white, and blue decorations. A grand auto parade took place on Friday, beginning at 7 p.m. in Guttenberg, about five miles north of the tunnel. Mayor Hague and Governor Miller led a procession that included motorcycle escorts, cars bearing the mayors of just about every city in northern New Jersey, fire engines, ambulances, and nearly one thousand floats representing various civic, educational, fraternal, business, and military organizations. The ten-mile-long parade, after winding its way as far south as Bayonne, terminated at Dickinson High School, at the end of the Twelfth Street Viaduct in Jersey City, where thousands of celebrants watched the grandest fireworks display in the city's history.

The following night, those not yet worn out by their walk under the river gathered at the high school at 8 p.m. for another, even greater display of pyrotechnics. This one lasted about two hours and included a replica,

executed in fireworks, of the aircraft *Spirit of St. Louis*, in which Charles
Lindberg made his epic solo voyage from America to France the previous May. At one end of Dickinson Field, a huge facsimile of the Statue of
Liberty burst into flame. Then the tiny winged replica, changing color as it
flew, sailed from a point near the statue several hundred feet across the field
until coming to rest beside a small-scale version of the Eiffel Tower. There
were ninety-five special explosive devices in all, including a brilliantly illuminated American flag suspended by a parachute five hundred feet in the
air. Although these displays made no specific reference to the tunnel opening, they reflected the general mood of patriotism associated with the tunnel's completion.

As the thunderous booms of the finale echoed in the Jersey City tunnel entrance and carried across the water to Manhattan, motorists lined up
and waited for the event that would top all others of the day. As midnight
approached, drivers began honking their horns, anxious to experience the
thrill of driving underneath the river. The first car through from New Jersey was a roadster driven by Charlotte Boettger, daughter of the New Jersey
tunnel commission chairman. Beside her sat her sister, Margot, and behind
in the rumble seat sat the temporarily displaced chauffer. Commissioner
Boettger, who had already been through the tunnel, was content to stand
and wave proudly as his daughter drove past the portal. John F. Boyle, Jr.,
son of New Jersey tunnel commissioner Boyle, was at the wheel of the second car through. His passengers were his father, Hudson County Detective
John McMahon, and Hudson County Press Club President Leo J. Hersh-
dorfer. The third vehicle and first "unofficial" car to go through from New
Jersey was driven by J. Frank Finn, a Jersey City attorney. Behind him came
thousands of cars, tucks, and limousines, spaced seventy-five feet apart
as they entered, in accordance with tunnel regulations. There were also
several motorcycles, including a few with sidecars carrying mothers with
babes in arms.

At the Manhattan tunnel entrance, there were hundreds of vehicles
waiting to enter but not nearly so many as in Jersey City. Commission
Chairman Dyer and Vice Chairman Bloomingdale rode through in the first
car admitted from New York, while Anna Holland and Gertrude Freeman,
widows of the first two chief engineers, rode in the second car. Behind
them followed a truck from the Bloomingdale Brothers department store,
the first commercial vehicle to go through from Manhattan.

During the first twenty-four hours of operation, 52,285 vehicles passed
through the tunnels, most of them passenger cars. Some of these were

repeats, however, driven by people who could not satisfy their curiosity with just one round trip. Inspector Cornelius Cahalane, consultant to the tunnel commissions in charge or organizing the tunnel police force, was standing in the New Jersey entrance plaza when a driver pulled up and asked which way to go to get to the nearest ferry. Cahalane told him the Lackawanna Railroad ferry was about a mile to the north and the Erie Railroad ferry just a few blocks to the south. He then asked the motorist why he did not just try the tunnel. "Tunnel," said the driver, "I've been through that seven times already."[5]

Air samples taken during the first twenty-four hours of operation revealed an hourly average of about one and one-half parts carbon monoxide to ten thousand parts of air, which was much less than the standard adopted. This was also about one-half the carbon monoxide content observed in the atmosphere of some streets in New York, as reported by the commissioner of health. Many motorists using the tunnel said that they found the air in the tunnel to be fresher than that on congested city boulevards.

Throughout the day on Sunday, the unbroken streams of traffic backed up to Newark in New Jersey and to the Brooklyn end of the Manhattan Bridge in New York. The normally heavy weekend traffic on the ferries was reduced by a half or more, although it was anticipated that once the novelty of driving under the river wore off and the work week began, the volume of vehicles in the tunnels would diminish. As predicted, the flow was much reduced on Monday, with 8,599 vehicles going through the south tube into New York and 9,127 going through the north tube to New Jersey. Commercial vehicles made up about 40 percent of the traffic. A check of the Pennsylvania Railroad ferries at Desbrosses Street showed that truck traffic was down slightly, while passenger-auto traffic was significantly decreased. But it would be a month or more before shipping companies and individuals adjusted to the availability of a new transportation option. Only then would the traffic counts and composition tell much about the project's potential financial success.[6]

By December 19, 1927, there were already strong indications that fears expressed by ferry operators prior to opening of the Holland Tunnel were being realized. An unofficial survey by the railroad corporations showed that the Pennsylvania Railroad had lost about 50 percent of its vehicular traffic, the Erie Railroad about 40 percent, the Lackawanna Railroad about 38 percent, the Central Railroad of New Jersey about 10 percent, and the New York Central Railroad about 10 percent. With revenue down sharply, railroad officials announced that they would have no choice but to raise rates or curtail service. A few days later, the Lackawanna Railroad took the latter course on its Christopher Street ferry.[7]

In May 1928, the Erie Railroad decided that instead of raising rates it would test the "elasticity" of demand by cutting rates from forty-five cents for roadsters, fifty cents for five-passenger cars, and one dollar for seven-passenger cars to twenty cents, thirty cents, and thirty-five cents, respectively, with an additional charge of four cents each for extra passengers.[8] In that month, the steadily rising average daily traffic in the tunnels had climbed to 24,384, and tunnel officials were predicting that if the trend continued, the tunnels would reach their maximum design capacity of 46,000 vehicles a day within a year.[9]

Part of the increase in traffic was due to a sharp rise in the number of buses using the tunnels. From a monthly count of 5,692 buses in January 1928, the number increased sharply to 20,233 in January 1929. Much of this increase was due to the creation of interstate bus lines, of which there were few in 1927. But some was due to tour operators' discovery that the tunnel had joined Chinatown and Grant's Tomb as a "must-see" part of the New York sightseeing experience. The ever-present Times Square tour barkers now had an addition to their pitch: "See the Holland Tunnel! Round trip through the new tubes now starting!"[10]

Some sightseers, of course, still preferred to do their own driving, and the tunnel police often found it difficult to keep traffic moving when drivers insisted on slowing down to get a good look at the ventilation slits along the curb, the fire boxes on the walls, and the state dividing line in the middle. Still others were content to loiter about the tunnel exits, just enjoying the sight of all the vehicles pouring out. One patrolman remarked, while pointing out a "little knot of earnest gazers" to a *New York Times* reporter, "They look like they expected something wonderful to pop out any minute." Another observer, watching the crowd as its members scrutinized each vehicle emerging from the portal, said, "Well, I guess they won't be satisfied until they see one of the cars coming out wreathed in seaweed."[11]

After a little more than one year of operation, the official traffic count for 1928 was 8,744,674, with about 78.4 percent passenger automobiles and motorcycles (21,102), about 2.6 percent buses, and the rest trucks.[12] According to a study conducted by the Hudson and Manhattan Railroad, that company lost 1,352,707 passengers in 1928 to bus lines operating through the tunnel. Despite an increase in the total number of railroad passengers traveling in and out of New York City in 1928, the ferries also continued to

see significant decreases in traffic. Not all of that was due to opening the
tunnel, because the Goethals Bridge and the Outerbridge Crossing opened
in June 1928, giving motorists an alternative means of getting from New Jersey to Staten Island. From 1927 to 1928, the total ferry traffic into or out of
the city dropped by 3,246,881 trips. The tunnel profit for 1928 was about $3.5
million, to be divided equally between the two states.[13] These trends continued in 1929, with a 25.5 percent increase in tunnel traffic and a 38 percent
increase in net revenue. Some of the revenue increase was due to a slight
reduction in operating costs, but there were many people who thought that
the operating costs should be much lower.

On January 7, 1930, Mark C. Kimberling, superintendent of Rahway
State Reformatory and former deputy superintendent of the New Jersey
State Police, testified before a legislative investigative committee headed
by New Jersey Republican State Senator Frank Dale Abell. Kimberling
had been retained to study operation of the Holland Tunnel in response to
claims that there were too many people on staff and that they were overpaid. The tunnel commissions employed nearly 550 people at this point,
none of them subject to civil-service regulations. Abell had begun his investigations the previous September by looking into the execution of contracts for construction and furnishing of state office buildings. His committee was now looking into the operating expenses of the tunnel commissions
and the Port Authority, although the results of the latter investigation were
not ready for presentation.[14] Kimberling charged that an annual waste of
$250,000 to $300,000 in tunnel revenue was due to exorbitant salaries, unnecessary employees, and duplication of effort. He recommended several
methods to cut operating procedures, but his most striking statements involved cuts in payroll.

Singstad, whose duties had lessened considerably in the two years since
the tunnel opened, received the greatest scrutiny. In comparing Singstad's
annual salary of $25,000 (well over $330,000 in 2010 dollars) to the $18,000
annual salary of the chief engineer of the New Jersey State Highway Department, Kimberling found that Singstad was "overpaid for the responsibilities connected with a complete and functioning organization." He
therefore recommended a cut in salary to a still generous $15,000 a year.
Other recommendations included a reduction in the salary of the chief of
tunnel police from $10,000 a year to $6,000 and replacement of two police
captain positions with lieutenants. He also urged abolishing twenty-six positions in the police department, the consulting physician position, one of
the two commission secretary-treasurer positions, fifteen positions in the

finance department, and ten positions in the tolls-collection department. In addition, he pointed out that unreasonably high salaries were paid to common laborers. "It will be noted," Kimberling stated, "that the salaries of all classes of [tunnel] employees, where there is a comparison with the civil service of the State of New Jersey, are very exorbitant."[15] And so they were.

The tunnel project had been used as a source of patronage since the first employees were hired, and the commissioners were not about to give up these staff positions without an argument. They responded predictably by stating that most of the cost-saving measures Kimberling suggested had been studied before the tunnel was open and found impracticable. As just one example, his suggestion that tolls be paid only at one end of the tunnel would result, they claimed, in increased ventilation needs and power costs.[16] Although they did not explain why they believed this to be true, it may be presumed that they thought more drivers would use the tunnel if they did not have to pay a toll, thus increasing the amount of vehicular exhaust in the tunnel.

The Abell committee's reports did little in the short term to change operating practices of either the tunnel commissions or the Port Authority. In the long term, the desire for greater operating efficiency and the need for another tunnel under the Hudson River at midtown resulted in a move toward consolidation of the commissions and the Port Authority into one body. This merger had long been advocated by Democrats in both states, who generally did not support the Republican-dominated tunnel commissions. Republicans, on the other hand, tended to favor a proposal by the tunnel commissioners to build up to five additional tunnels.

By the beginning of 1930, New York Governor Franklin D. Roosevelt had become a very strong supporter of consolidation, which could be achieved by adding members of the two tunnel commissions to the membership of the Port Authority. He believed that such a move would combine the operational advantages of a united engineering and technical staff with the financial advantage of allowing the Port Authority to pay for new projects by issuing its own bonds, secured by operating revenue. In a speech at the Holland Society of New York's annual dinner, held in New York City's Hotel Astor on January 16, 1930, Roosevelt called for a "wedding" of the joint tunnel commissions with the Port Authority. In so doing, he appropriated the metaphor first used by Al Smith in 1919 to encourage bistate agreement on construction of a vehicular tunnel and agreement on creation of a port authority. Now, however, the very profitable tunnel could be seen not so much as a "ring" symbolizing mutual commitment but as a dowry.[17]

In 1930, the New York and New Jersey legislatures each passed five bills that, as a group, abolished the tunnel commissions and expanded the membership of the new Port Authority to twelve members, equally apportioned between the two states. Of the six tunnel commissioners appointed to the new body, only George Dyer and A. J. Shamberg had had anything to do with building the Holland Tunnel. The legislation also turned over tunnel operation to the Port Authority and authorized expenditures for a survey of a midtown Hudson River tunnel (or tunnels).[18]

In 1931, the next steps were taken when the two states authorized transfer of future tunnel revenues to the Port Authority upon its issuance of $50,000 in bonds to reimburse the states for their investment, and permitted the Port Authority to pool surplus revenue for the establishment of a General Reserve Fund to protect outstanding and future bond issues. The legislatures also authorized construction of the midtown Hudson River tunnel.[19]

On March 19, 1931, Holland Tunnel traffic reached thirty-five million, far ahead of projections made when it opened. Preliminary data indicated that despite the deepening economic depression, traffic in 1931 was already about 7 percent heavier than in the previous year. Two days later, checks totaling approximately $50 million were handed by the Port Authority to representatives of the states of New York and New Jersey as reimbursement for their investment in the tunnel. Since the total cost of the project, including interest charges on the money used for construction and similar items, came to about $54 million, the states had not quite yet recovered all of the cost, but there was no doubt that they would do so ahead of schedule.[20]

March 1931 thus marks an important point in the history of the Holland Tunnel. After nearly two and a half years of operation, the tunnel had proved that long vehicular tunnels could be safely ventilated. The payments made by the Port Authority to the two states also served as proof of the project's financial success, while the traffic counts demonstrated the need for an additional tunnel (or tunnels) under the Hudson River.

The new midtown tunnel would be built as part of the Port Authority's comprehensive plan, which also included a massive, sixteen-story inland freight terminal, to be constructed on the block bordered by Eighth and Ninth Avenues and West Fifteenth and West Sixteenth Streets, placing it almost equidistant from the Holland and midtown tunnels. As opposed to the smaller inland freight terminal previously used by the railroads, the new terminal would be a "union" facility, designated for use by all railroads and shippers. The Port Authority awarded a contract for the excavation and foundation work on Inland Terminal No. 1, designed for handling less-than-carload-lot freight, at the beginning of April 1931. It was supposed to be the first of three Manhattan freight terminals to be built by the Port Authority.

Two other components of the Port Authority's comprehensive plan were completed by the end of the year. The George Washington Bridge between Fort Lee, New Jersey, and 178th Street in Manhattan opened on October 25, 1931, and the Bayonne Bridge to Staten Island opened November 15, 1931. These links joined the Goethals Bridge and Outerbridge Crossing, both completed in 1928, as part of the area-wide transportation system managed by the Port Authority. At the end of 1931, therefore, the future of vehicular transportation between New York and New Jersey seemed promising. Problems, however, soon arose.

The three Port Authority bridges connecting New Jersey with Staten Island (Bayonne, Goethals, and Outerbridge Crossing) were money losers and continued to be so for decades to come. And the railroads reneged on their commitments to use Inland Terminal No. 1 to its full capacity after it opened in September 1932 and instead maintained their existing freight-distribution systems. As a result, the terminal was not profitable, and the other inland freight terminals were never built.[21]

Another problem involved the lack of proper highways in Manhattan and New Jersey to handle the traffic generated by the Holland Tunnel. At noon on November 13, 1930, Manhattan Borough President Julius Miller led a ceremony marking completion of the first section of the West Side Elevated Highway, one of the world's first urban freeways. This initial segment, completed at a cost of $6.5 million, ran from Canal Street to Twenty-Second Street. Eventually, the highway would extend to Seventy-Second Street and then over the tracks of the New York Central Railroad and along the waterfront to the Bronx. But it was not until February 1937 that completion of the portion between Forty-Sixth Street and Fifty-Ninth Street allowed traffic to flow all the way from Canal Street to Seventy-Second Street.

City crews erected a speaker's stand and grandstand in Canal Street just steps away from where the tunnel groundbreaking ceremonies had been held years before. At the appointed hour, a large crowd gathered to hear various officials praise each other for their vision in bringing about the epic achievement of the highway. Speaking into microphones of the National Broadcasting Company and radio station WNYC, Lincoln Cromwell, vice president of the Automobile Merchants' Association, summed up the expectations of many people when he said that the highway would "lessen the cost of business by speeding traffic."[22]

After the speeches had all been made and the ceremonial ribbon cut, the officials and dignitaries climbed into cars and then ascended the steep Canal Street ramp, followed by the usual police motorcycles, fire trucks, ambulances, and marching military units. This parade also had something extra. Because it began so near the New York City Sanitation Department garage located next to the tunnel ventilation building, the department's employees jumped into their trash trucks, snowplows, and other vehicles and joined the parade. The celebrants then drove or marched for one and a half miles past the decorated third-floor windows of merchants along West Street before descending the equally steep and temporary wood ramp at Twenty-Third Street. The highway was then closed until November 17 because it was not yet finished.

The Port Authority had opposed the highway because it did not fit in with their plans to direct truck traffic to Inland Freight Terminal No. 1. The Women's League for the Protection of Riverside Park also opposed the project because it did not want to see truck traffic routed through the park. As a consequence, the New York City banned trucks from the highway, even though they constituted approximately 16.8 percent of the traffic using the tunnel.

Although the highway had been hailed as the beginning of "a new system of express highways," it was poorly designed. The road surface was slippery in wet weather, drainage was poor, the steep and poorly placed ramps were difficult to negotiate, the roadway was too narrow, and its capacity was so limited that it quickly became congested. In order to lessen the expense of building condemnation, sharp curves were included in the alignment, thus reducing the speed at which vehicles could travel.

In New Jersey, the last three-mile segment of the Route 1 Extension, otherwise known as the Jersey City–Newark Elevated Express Highway (and now known as the Pulaski Skyway), was not completed until November 1932. Described by Thomas H. MacDonald, chief of the Federal Bureau of Public Roads, as the "the greatest highway project in United States today," the road suffered from some of the same design problems as the West Side Elevated Highway: the road surface was slippery in wet weather, the ramps were difficult to negotiate, the roadway was too narrow, and the capacity was limited. It was not long before trucks were also banned on this highway.[23]

Engineers had not yet learned how to properly design superhighways, but they learned from their mistakes, at least in regard to how to make highways safer and to improve their capacity. The same could be said in regard to tunnels: lessons learned on the Holland Tunnel were applied to the design of virtually all the vehicular tunnels that came after it.

Plans for the midtown Hudson River vehicular tunnel, which would represent a slight improvement on the Holland Tunnel design, with wider roadways, were put on hold in 1931 because of a weakening of the bond market. It was not until 1933 that Robert Moses, chairman of the New York State Emergency Public Works Commission and head of many other public entities, obtained the necessary funds for the Port Authority to begin construction. The Port Authority's chief engineer at that time was Othmar Ammann, a man primarily known for his expertise in bridge design. Ammann hired Singstad as a consulting engineer, and the chief designer of the Holland Tunnel thus became the chief designer of the first tube of the midtown tunnel. In April 1937, the Port Authority decided to rename the Midtown Hudson River Tunnel the "Lincoln Tunnel" in order to associate the structure with one of the nation's greatest presidents (as it had done with the George Washington Bridge) and to differentiate it from a planned midtown tunnel under the East River to Queens.[24] The first tube of the Lincoln Tunnel opened in December 1937, and the second tube, also designed by Singstad, opened in February 1945.

On April 8, 1935, New York Governor Herbert Lehman signed a bill creating the Queens-Midtown Tunnel Authority. Singstad was subsequently appointed chief engineer and given the difficult task of building the new tunnel under the bed of the East River. Singstad retained his position when the Port Authority was re-created in January 1936, as the New York City Tunnel Authority, with authorization to build both the Queens-Midtown Tunnel and a two-mile-long tunnel between Brooklyn and the Battery in Lower Manhattan. Opened in November 1940, the Queens-Midtown Tunnel cost $54 million, about $4 million under the original estimate. Singstad planned and oversaw initial construction of the Brooklyn-Battery Tunnel, on which construction began in October 1940. Work was suspended in 1943, however, due to funding and material and labor shortages brought about by World War II, and Singstad resigned his position as chief engineer in 1945.

In 1946, the New York City Tunnel Authority merged with the Triborough Bridge Authority, creating the Triborough Bridge and Tunnel Authority. That agency completed the Brooklyn-Battery Tunnel, according to specifications imposed by Triborough Bridge and Tunnel Authority Chairman Robert Moses. The basic design was Singstad's, however, and like the Lincoln and the Queens-Midtown tunnels, it marked an improvement on the Holland Tunnel design while also maintaining the most important design features of the earlier structure.[25]

Prior to completion of the Brooklyn-Battery Tunnel, Singstad began to envision a cross-Manhattan vehicular tunnel as a way to link the Lincoln and Queens-Midtown tunnels. This project, he hoped, would relieve some of the traffic congestion caused by the earlier two projects. But the cross-Manhattan tunnel would forever remain an unrealized dream.

The ventilation system that Singstad first designed for the Holland Tunnel and then used again in his other projects was his greatest legacy. It worked well in keeping the air within the tunnel safe under normal operating conditions. But it was not perfect. There had been 192 fires in the Holland Tunnel in its first year of operation, and each year thereafter, there continued to be multiple fires. For the most part, these were very minor and easily extinguished. But John O'Rourke had been correct in his criticism of the tunnel fire tests conducted prior to opening in 1927. Extinguishing the fire from one burning vehicle is not the same thing as putting out the fire from several burning vehicles, particularly if some of them are filled with explosive or inflammable materials.

On March 19, 1932, John Sacco, twenty-one, of Deal, New Jersey, was behind the wheel of a car racing through the westbound tube of the Holland Tunnel at about sixty miles an hour when he swerved from the fast left-hand lane into the slow lane to pass another car. Apparently, he did not see what was ahead of him until it was too late. His car struck the rear end of a National Biscuit Company truck, killing Sacco and seriously injuring his two passengers. He thus became the first person to die in the tunnel since its completion.[26] There was no fire in this instance, and emergency crews cleared the wreckage of Sacco's light coupe, scattered over both lanes of traffic, within twenty minutes. But what if the truck he hit had been carrying something more combustible than cookies? What if there had been a fuel leak and fire, igniting not just Sacco's car but also the truck and other vehicles? How would the ventilation system and emergency procedures work then?

12

Fires, Blasts Rip Holland Tunnel

Fires, Blasts Rip Holland Tunnel
   —subhead of article in the *New York Times*, May 14, 1949

IT WAS JUST past 8 a.m. on Friday the thirteenth in May 1949, a regular workday for recently employed truck driver Edward Tyndall, a thirty-nine-year-old resident of Weehawken, New Jersey. Tyndall pulled his sixteen-ton tractor-trailer rig out of the Boyce Motor Lines terminal at 241 Colden Street in Jersey City and headed for Pier 2, Thirty-Third Street, Brooklyn. Tyndall received no information about what he was transporting inside the thirty-one-foot-long enclosed trailer. He was not given a bill of lading, and the trailer had no mark or placard indicating its contents. He was simply instructed to deliver the trailer to the loading dock of Farrell Lines Incorporated. There the trailer's contents were to be loaded onto the SS *African Glade* for delivery to the European Chemical Company in South Africa.[1]

What Tyndall was not told was that his trailer was loaded with 48,536 pounds (twenty-four tons) of carbon disulfide, stored in eighty steel drums, each containing fifty-five gallons of this highly inflammable, extremely toxic chemical. Manufactured by the J. T. Baker Company at its Taylor Chemical Division in Cascade Mills, New York, carbon disulfide was commonly used as a solvent for fats and oils, in the manufacture of rubber and rayon, as a fumigant, and in the manufacture of explosives. Its ignition point is lower than that of gasoline. The heat from a steam pipe, a light bulb, or even from a metal roof heated by the sun can be enough to ignite its vapors. When burning, it produces deadly carbon monoxide and sulfur dioxide.[2]

It was likely that Tyndall also did not know he was subject to a regulation of the Interstate Commerce Commission (ICC) (18 U.S.C. Section 835) providing that drivers of motor vehicles transporting inflammables or explosives had to avoid, by prearrangement of routes, driving into, over, or through congested thoroughfares, places where crowds were assembled, streetcar tracks, viaducts, dangerous crossings, and tunnels. The regulation did not prohibit the transportation of inflammable or explosive materials, nor did it necessarily prohibit them from being transported through tunnels. It simply required that the route of travel should be selected, in advance, to avoid tunnels and other points of congestion, "as far as practicable." A 1935 federal case, *Sproles v. Binford* (286 U.S. 374), established that the route selected should be the "shortest" practicable route.[3]

Boyce Motor Lines determined, as it had on two previous occasions, that the shortest practicable route for transporting carbon disulfide between Jersey City and Brooklyn was through the Holland Tunnel, the route Tyndall was directed to take. Each time, however, Boyce Motor Lines neglected to comply with ICC and Port Authority rules regarding how the chemical was to be contained and placarded for transport.

Boyce Motor Lines was not alone in its disregard for regulations designed to protect the public from the hazards involved in transportation of explosives and other dangerous substances. As later investigation proved, it was common for freight companies to send potentially hazardous materials through the tunnel. After all, the main reason the tunnel was built was for the expeditious movement of freight between Jersey City and the boroughs of New York City. It was logical for companies to use that route. Moreover, it was relatively easy to slip hazardous cargo past the police. Since the first of the year, for example, only about one thousand suspicious-looking trucks had been stopped and examined at the entrances to the Holland and Lincoln Tunnels, and only about fifteen of those had been turned back.[4] Given the Port Authority's lax enforcement of regulations, many companies saw the tunnel as the most cost-effective option for transporting their freight across the Hudson River.

Prior to Tyndall's trip on May 13, there had been no catastrophic accidents, but there had been events that should have served as warnings. Minor fires were not uncommon; more than fifty a year had occurred since the tunnel opened in 1927. On one occasion, an auxiliary fuel tank fell from a truck and ruptured, spewing gasoline for several hundred feet along the roadway before the tank separated from the vehicle. The vapors did not ignite, but traffic was held up for hours while the fuel was washed into the sumps, which were filled and emptied three times before traffic could be allowed to move once again through the tube.[5] And just a few years before Tyndall's incident, a much more serious accident occurred. At 5:40 in the evening of January 2, 1946, Albert Cook, a forty-four-year old driver

from the Bronx, was behind the wheel of a tractor-trailer with a twelve-
to fourteen-ton load of shoe polish and chewing gum. He was midway
through the westbound tube of the tunnel when the truck in front of him
stopped suddenly, forcing him to do the same. His trailer swerved sharply,
first toward the left-hand "express" lane and then back into the "slow" lane,
as he fought to bring it under control. The tractor remained upright, but
the trailer flipped over, snapping the coupling from the tractor. The truck's
two thirty-five gallon tanks of gasoline exploded, setting both the cab and
trailer on fire. The heat not only melted the load from the trailer, making a gummy mess that covered the roadway, but also melted tiles in the
tunnel walls.

Two of the Port Authority's Jersey City emergency crews, together with
a Jersey City Fire Department engine company, battled the blaze from their
end, while two New York City–based Port Authority crews labored to clear
vehicles from their end. Approximately one hundred vehicles had entered
the westbound tube when the accident occurred. Some automobiles were
able to turn around and exit the way they came in, but most of the trucks
had to be backed or towed out. As frustrated drivers honked their horns,
approximately thirty police officers did their best to unsnarl the massive
knot of vehicles that developed in the vicinity of the New York entrance
plaza. Drivers were instructed to divert to the Lincoln Tunnel, the George
Washington Bridge, or the ferries.

Around 8:00 p.m. Jersey City crews succeeded in hauling out the tractor
that Cook had been driving, but it took another hour before the heavily
damaged trailer could be dragged out on its side. At 9:24 p.m., four hours
after the accident occurred, one lane of the westbound tube was finally
cleared of debris, allowing traffic to pass through to Jersey City. Cook, the
only person reported hurt, was treated at Jersey City Medical Center for a
slight leg injury and released.[6]

The potential for serious accidents to take place in the tunnel was increased by the tendency of drivers, sometimes at the urging of Port Authority police officers, to disregard the rules requiring a minimum distance
of seventy-five feet between vehicles and a speed of no more than thirty
miles per hour.[7] When traffic levels were light, both the distance between
vehicles and the average speed tended to increase, but when traffic was
heavy, the average speed and the distances between vehicles tended to decrease as they bunched up. The accident of 1946 was probably due, in part,
to Cook's having followed the vehicle in front of him too closely.

On the morning of May 13, 1949, traffic inside and outside the tunnel moved slowly. Normal weekday rush-hour traffic ranged from fourteen to eighteen hundred vehicles per hour in each tube, but occasionally the total load in both tubes reached four thousand vehicles per hour. Lines of cars, trucks, and buses often stretched for miles through city streets approaching the entrance plazas. It took Tyndall more than half an hour to negotiate the approximately two-mile path to the Jersey City portal, and he entered the eastbound tube at about 8:45 a.m. He was approximately twenty-five hundred feet in, driving in the right-hand lane, when he stopped momentarily for a traffic signal. He then heard what he later described as a "loud boom." Looking in the rearview mirror, all he could see were flames coming from his trailer. After swerving into the left lane, around a truck in front of him, and then back into the right lane, Tyndall opened the door and leapt out, leaving the truck to roll back into the left lane. Running for his life toward the Manhattan exit, he saw a truck up ahead that was stopped, and he climbed into the cab. The driver then sped off to exit the tunnel in New York. After alighting from the truck at the exit plaza, Tyndall spoke briefly with Port Authority police and then was taken to Beckman Hospital for examination.[8]

Frederick Ridents, of Flushing, Queens, the driver of an empty truck directly in front of Tyndall, heard what he thought was a tire blowing out. Looking back, he saw Tyndall's rig on fire, with one explosion rapidly following another. Blocked by traffic up ahead, he could not drive away from the danger. "I jumped from my truck and I looked back. Trucks were going up in flames. That was enough for me. I wrapped my handkerchief around my face and began running. I must have run one mile through automobiles and trucks to New York," he later told a reporter for the *New York Times*.[9]

Eric Hill, of Jersey City, was driving a truck for the Baldwin Oil Company loaded with paint, turpentine, linseed oil, and other paint supplies. "I was two trucks ahead of the big trailer truck when I heard an explosion. I stopped, turned around, and saw flames and smoke." Hill and his assistant, twenty-one-year old Joseph Dowgials, jumped from their truck and started running toward New York. Hill caught a ride in a car, and Dowgials hopped onto another truck. "All the way through the tunnel we could hear those explosions," he said.[10] Drivers of other cars and trucks also leapt from their vehicles and began running toward the exit, although some were reported to have been "blown out of the seats of their machines."[11] Edward Daly, of Jersey City, said that the explosion blew the doors off his truck, which then caught fire. "That was enough for me," said Daly. "I got the hell out."[12]

There were two lighting circuits in each tube of the Holland Tunnel, with alternate lights on each one, designed so that failure of one circuit would leave half the lights on. But the system was not designed to handle a fire of the magnitude that developed. Soon all the lights went out, and the tube began filling with a yellowish-black cloud of smoke and toxic vapors, periodically pierced by flashes of light from exploding, burning vehicles and their cargos. John McMahon, a twenty-nine-year-old Port Authority patrolman from Newark, was stationed in a tiny booth on the sidewalk near the point where Tyndall's truck came to rest. Ignoring the threat to his own safety, he remained in place, shouting to drivers to link hands and make their way in the near darkness toward the New Jersey exit.

The first emergency signal transmitted to the Port Authority Police at Vestry and Varick Streets in Manhattan came at 8:48 a.m. from a patrolman stationed east of the burning vehicles. He saw a truck stopped about one hundred feet west of his position and transmitted a signal for "trouble or breakdown." The supervisor on duty recorded this signal and switched the south-tube traffic lights to amber, which was intended to slow but not stop traffic. As the patrolman ran toward the stalled truck, he heard a loud blast and then met two men running toward him. After guiding them through a connecting passageway to the north tube, he returned to the south tube to find a collapsed Port Authority patrolman, whom he also assisted to the north tube.

Meanwhile, a truck driver ordered by a patrolman to drive his vehicle out of the tunnel staggered into the New York tunnel office, reporting that he had been "gassed," and then collapsed. This was observed by George Schiffer, a forty-five-year-old Port Authority sergeant from Brooklyn, who, according to amber-signal procedure, was standing by with his two-man emergency team. About that time, at 8:56 a.m., the control room received the first fire alarm. Schiffer jumped into his specially outfitted jeep, the other two men jumped into an emergency wrecker with steering controls at either end, and both vehicles headed into the tunnel. On the way, Schiffer stopped to pick up two patrolmen and then continued to a point where he could see the fire. "When we got there, the place was jammed with burning trucks. We couldn't get a glimpse of how many trucks were on fire because of the heavy smoke," Schiffer reported.[13] After putting on gas masks, the crew pulled a hose to a standpipe about seventy-five feet east of the fire. They connected the hose and then advanced toward the large ball of fire ahead of them, preceded by the spray emanating from the hose's fog nozzle. They soon extinguished the fire in the first truck, containing meat, and the second truck, containing bleach, and were spraying the third truck, loaded with paint supplies (Hill's), when they noticed flaming liquid running down the gutters on either side of the roadway. Recognizing the need for greater assistance, Schiffer ran to the nearest telephone and called for more help.

The first "red" alarm may have been sent by Joseph Lushkin, a twenty-seven-year-old Port Authority patrolman from Brooklyn, whose station was a short distance east of where Tyndall's truck had come to rest. Lushkin saw a gray cloud of smoke coming toward him but thought little of it because, as he later reported, "we get a lot of it."[14] Not recognizing the danger at first, he walked right into the cloud, which smelled of ammonia. He returned to his station and telephoned to the supervisor on duty to pump more air. But according to established fire-emergency procedures, the blower fans in any one section of tunnel with a fire were to be shut off while the exhaust fans were left running, and the exhaust fans in adjacent sections (seven sections in all) were to be shut off while the blower fans were left running. This procedure was designed for small localized fires, however, not for a large fire quickly spreading through multiple sections. Apparently, lack of knowledge in the control room of the extent of the fire caused delay in figuring out which combination of exhaust and blower units should be used.

As Lushkin asked for more air, other patrolmen came running up to him, shouting, "Hit the reds!"—meaning "Sound the fire alarm!" About this time, he saw flames and heard explosions. With the alarm turned in, he and another patrolman began running toward the fire. Lushkin, however, was soon overcome by smoke and had to be evacuated to St. Vincent's Hospital.[15]

The fire alarm resulted in all traffic signals west of Lushkin's station turning to red, halting movement into the eastbound tube. Three buses filled with about 120 schoolchildren were stopped just short of the Jersey City entrance portal, and the children were ordered off. But many other buses, automobiles, and trucks had already entered the tube and were jammed up inside.

First responders entering from Jersey City were partially blocked by more than 125 vehicles that had been abandoned by their drivers in both lanes. They therefore had to carry hoses all the way from the entrance, past the parked vehicles, to standpipes close to the fire. These standpipes were located every 120 feet along the length of the tunnel. When the tunnel first opened in 1927, hoses were connected to the standpipes and left in place, but the necessity of frequent inspection and replacement resulted in their removal. Now the firefighters battling the blaze would pay the price for that change.[16]

At 9:05 a.m., the Jersey City Fire Department received the first accident report and dispatched a rescue unit: pumper and emergency truck. As the first crews entered from the New Jersey side, McMahon left his post at the security booth near the fire and moved west to help the firefighters and a few courageous drivers who stayed behind to remove abandoned vehicles. Like so many others who were on the scene when the accident occurred, he was overcome by smoke and fumes and had to be evacuated to Bellevue Hospital.

Jersey City Fire Chief Frank J. Ertle, sixty-four, entered the south tube soon after the first crews. After going about twenty feet, he detected the strong smell of sulfur dioxide. He returned to the surface to put in a second alarm and to order that all available oxygen masks be brought to the tunnel entrance. After being equipped with masks and rolled lengths of hose and tools, additional Jersey City firefighters entered the tube and advanced between the lines of abandoned vehicles until they reached the first burning trucks. After partially extinguishing flames in the first two vehicles encountered, they moved forward about two hundred feet, while others worked to clear trucks out of the tube behind them.

After Lushkin turned in the fire alarm, the lights east of the alarm point flashed amber, signaling that motorists should drive their vehicles out in the right lane, thus leaving the left lane open to emergency vehicles entering from Manhattan. Except for the four trucks immediately in front of Tyndall's truck—Ridents's empty rig, Hill's truck carrying paint supplies, the truck loaded with bottles of bleach, and the truck filled with meat —all vehicles east of the fire managed to exit the tube under their own power. This enabled New York emergency crews to drive in as close to the flames as the intense heat would allow. About forty-five minutes after the fire started, they were able to haul out their first truck, the one containing meat, but the others were too damaged to be moved. The toxic fumes produced by burning cargos in these trucks greatly complicated the firefighting and made the work of emergency crews from New York particularly hazardous.

Shortly after Sergeant Schiffer observed the first truck driver stumble into the New York office to report the fire, other drivers emerging from the tunnel also reported to a New York City policeman stationed at the exit plaza that they had been "gassed." This officer relayed the information to his dispatcher, who sent two police emergency squads, two patrol cars, and an ambulance to the tunnel at 8:59 a.m. But it was not until 9:12 a.m. that the police put in a call to the New York City Fire Department (FDNY), reporting that "a drum of chemicals had fallen from a truck, . . . and fumes were filling the tube."[17] This report of a drum falling from the truck was later cited as the cause of the accident, but the report was mere speculation. No one was in a position at that time to know why the accident had happened. And given that the trailer was fully enclosed, it is unlikely that any of the drums could easily have fallen out. Later in the day, New York City Chief Fire Marshall William P. Murphy stated that the first explosion was due to a chemical reaction caused by the change in atmospheric pressure in the tunnel. For the moment, however, those fighting the blaze cared little about the cause of the fire; they were preoccupied with putting it out, without losing their lives in the process.

At 9:13 a.m., New York City Rescue Company Number 1 and Battalion Chief John Heaney of the Fifth Battalion were dispatched to a rallying point at West and Spring Streets near the entrance to the westbound tube, which was closed to traffic at 9:20 a.m. after it began filling with smoke seeping in from the eastbound tube. Heaney led the rescue squad into the tunnel and down to a point approximately opposite the fire in the eastbound tube, where they crossed over via a connecting shaft. The scene, lighted by handheld emergency lamps, was one of complete chaos. Temperatures in the immediate vicinity of the fire, later estimated by one expert to have reached four thousand degrees at the height of the conflagration, were intense. Partially blinded by their masks, the firefighters struggled to see through the thick smoke. Correctly assessing the severity of the fire, Heaney called the tunnel switchboard operator and asked that a full first-alarm assignment be relayed to FDNY.

After receiving news of the fire's severity, William J. Hennessey, FDNY assistant chief of staff and operation, notified headquarters that all available oxygen helmets were needed. Chief Murphy then put out a call to rescue-squad companies in the Bronx, Queens, and Brooklyn to send their entire supply of helmets. One rescue truck sped through the Lincoln Tunnel to deliver helmets and additional firefighters to the Jersey City Fire Department. An FDNY ambulance equipped with oxygen masks was also dispatched to the New York end of the tunnel, while another went to the New Jersey end.[18] In addition, four Consolidated Edison of New York emergency trucks were sent to the scene with inhalers.

One of the first ambulances to reach the west entrance of the eastbound tube was from Christ Hospital, Jersey City. Dr. William Raymond, the physician on board, later reported what happened when they arrived: "Nobody seemed to know just what the situation was or what was going on. Clouds of smoke were pouring from the mouth of the tunnel, and shortly after we arrived we heard several muffled explosions from inside the tunnel, coming in rapid succession. We continued to hear these explosions coming in bursts of two and three, each burst spaced several minutes apart." Raymond administered oxygen to firefighters overcome by fumes and smoke as they were carried out of the tunnel. He noticed that the firefighters who used carbon-filled civil-defense-type canisters seemed to be worse off than those using chemical canisters. Most, however, soon recovered and insisted on going back into the tunnel, only to be carried out a second time in much poorer condition.[19]

Inside the tunnel, many firefighters and rescue personnel were finding that the only way they could breathe was by getting on their knees and sticking their faces next to the fresh-air inlets located slightly above the level of the road surface along the curb, spaced about fifteen feet apart. The flow of air through the ventilation system was increased shortly after Raymond's ambulance arrived at the Jersey City entrance, and firefighters could then see and breathe better without their masks.

Increasing air flow solved one problem but made another worse. The intense heat from the exhaust gases began to damage the fans in the Jersey City ventilation building. First one fan and then another failed, while paint in the fan room blistered and scaled off. Soon two of the six fans ventilating the section where the fire was most intense were out of service, and a third was about to go. Port Authority maintenance personnel, seeing that the entire system was in jeopardy and ignoring the heat, grabbed hoses and began spraying water on the fans still in operation.

The increased air flow cleared enough of the smoke from inside the tube to allow emergency personnel to observe extensive damage to the overhead "ceiling" of the tunnel. Huge chunks of concrete, five inches in thickness, had fallen down, sometimes striking emergency personnel, while other sections hung by remnants of the reinforcing steel mesh. The barrier between the vehicular tube and the exhaust-air tube was breached for a distance of approximately two hundred feet, allowing flames to lap directly against the concrete inner lining of the cast-iron outer shell. This layer began to crack and fall away as the intensity of the fire increased. When word of the tube's condition reached the control center, the fireboat *John J. Harvey* and the fire tender *Smoke* were sent to a spot on the Hudson River directly over the tunnel to search for bubbles or other signs of an air leak, which would indicate that the fire was eating through the outer shell.[20]

The most obvious external indications of the fire were the plumes of smoke emanating from the tunnel entrance and exit and a thick column of smoke coming from the New Jersey land ventilation building that could be seen for miles. The smoke, numerous emergency sirens, and radio reports soon drew hundreds of onlookers to both ends of the tunnel, where crowds of the curious complicated emergency operations by getting in the way. One observer, Peter Kalakowski, a forty-nine-year-old Jersey City resident, was hit by a truck at the Jersey entrance and had to be rushed to the Jersey City Medical Center with possible fractures to both legs.[21]

Other, more distant indications of the fire were caused by disruptions of telephone, telegraph, and radio connections. On the south side of the tube between the circular outer shell and the vertical inner wall, ducts carried five communication cables. After the fire destroyed the vertical wall, the cables were exposed, and they too were destroyed around 9:20 a.m., knocking out about half the long-distance telephone circuits to points west and south of the Hudson River. One of these cables was a nine-hundred-circuit trunk line of the American Telephone and Telegraph Company, which carried signals from central phone offices in northern New Jersey to the company's main office in New York. A spokesman later said that the disruption was the worst ever experienced by the company. Also involved was a coaxial cable capable of carrying either telephone or television signals, although it was not in use for television at the time. Other cables carried circuits used by press associations, radio networks, and business and financial firms, and two hundred lines of the Western Union Telegraph Company.[22]

Area transportation patterns were also severely affected, and, as in 1946, drivers were detoured to the Lincoln Tunnel, the George Washington Bridge, and the ferries, all of which were soon loaded to capacity. The Lackawanna Railroad placed another boat into service on its Barclay Street ferry and, along with the Christopher, Chambers, and Liberty Street ferries, decreased the time that each boat spent unloading and loading vehicles.

Non–Port Authority emergency vehicles struggled against tied-up traffic to reach the scene, and each unit was sorely needed. So many of the first firefighters to go into the tube were put out of action by smoke, fumes, and falling debris that the need for greater manpower and equipment soon became obvious. At 11:19, a second alarm was transmitted to FDNY headquarters, and for the first time in its history, the department was forced to use four rescue companies on a single fire. Eventually, more than forty firefighters from outlying boroughs of New York were brought in to help Manhattan-based units.

As additional help began arriving, Father John Shields of St. Alphonsus Roman Catholic Church, 308 West Broadway, grabbed a seat on a Port Authority jeep and rode into the tunnel to minister to the injured. "It was a hot box," he later told a reporter, "and it seemed that trucks and automobiles were caught in one gigantic pocket. The fire, smoke and the smell were intense." Yet despite the exploding containers of chemicals, debris, and general disorder, Shields noticed that the firefighters were working with "amazing efficiency."[23]

By the time the second alarm was turned in on the New York side, the Jersey City firefighters and rescue squads had succeeded in removing all undamaged vehicles from their end of the tube. This allowed two 1,250-gallon-per-minute pumpers to be driven right up to the fire, with each laying hose as it approached. These units were connected to a third pumper, which was stationed at the entrance and connected to a standard city fire hydrant. This was done to augment the tunnel's standpipe system and to serve as a backup in case the tunnel's system failed. In all, the Jersey City Fire Department sent six pumpers, three emergency trucks, one rescue unit, and one lighting truck to the scene.[24] The Jersey City firefighters focused their efforts on a group of five trucks located approximately 350 feet to the rear of the group that included Tyndall's truck. This westernmost group included, from west to east, trucks loaded with wood barrels, rolls of newsprint, eight hundred cases of tomato juice, groceries and clams, and wax.

At 12:40 p.m., ten reporters, a driver, and a Port Authority representative crammed into an eight-passenger car at Watts Street and drove into the westbound tube. Following the same path as that of Battalion Chief Heaney and his team nearly three and a half hours before, they crossed through a connecting passageway and then walked down to the scene of the fire. There they found firefighters spraying the melted trucks and still-smoldering debris, while keeping an eye on the chucks of concrete and long strips of steel mesh hanging over their heads. One reporter thought that the tunnel "gave every indication of having been struck from the inside by a bomb." Two Marine Corps majors from the Third Naval District, who had been in the tunnel, stated that the destruction was worse than what they had observed in bombed-out cities during the Second World War.[25]

By about 1:00 p.m., approximately three hours after the fire began, it was mostly out. The Port Authority opened the westbound tube, now free of smoke, to emergency traffic at 2:05 p.m. and opened it to all traffic, with one lane for eastbound and one lane for westbound vehicles, shortly thereafter. At first, traffic was light, but as word spread that the Holland Tunnel was back in partial operation, the lines began to back up.

As the north tube reopened, the effort in the south tube quickly switched from firefighting to cleanup, and FDNY searchlight trucks were brought in to provide illumination. Despite lingering heat from the walls and piles of wreckage, and continued poor visibility in the murky gloom, emergency crews began removing wreckage and hauling out the burned trucks and trailers. It took approximately sixteen hours to remove all of them, with the task greatly complicated by a thick layer of broken concrete and tile on the road surface and slick pools of water and sludge all around. The rolls of newsprint were a particular problem; they continued to burn long after the truck they were in had been taken out the New Jersey side. Four dump trucks were brought in from Idlewild and LaGuardia Airports, along with scoop shovels, trucks, and other heavy equipment. Shortly after 5:00 p.m., portable floodlights from Newark Airport were also brought in from the Jersey City side.

At 6:20 p.m. (or maybe 6:34 p.m. or 6:50 p.m.; accounts differ), long after it was assumed that the fire had been extinguished, spilled kerosene and alcohol leaking from the Baldwin Oil Company truck reignited, causing a series of muffled explosions in the gutters and drains. Three additional drums of carbon disulfide also blew. Crews that had come in to clear away debris were driven back as firefighters rushed in to fight this new threat using more than twenty five-gallon foam extinguishers. They then set up a foam generator to mix thirty cans of powder with water and began covering the area with heavy foam. About 7:30 p.m., an additional drum of carbon disulfide blew, without doing much damage. Finally, at about 8:15 p.m., the wrecking crews succeeded in taking out eight drums of kerosene and alcohol from the Baldwin Oil Company truck, thus removing the last remaining threat of further explosion.[26]

The remains of the burned trucks and trailers were particularly hard to separate from the ankle-deep debris, and most of them had had their wheels entirely blown off. The trailer carrying cases of canned tomato juice had to be cut away from its tractor with acetylene torches, and around 11:00 p.m., the heat from the torches burst the cans, which sprayed the tunnel walls and salvage crews with juice as the trailer was hauled out the Jersey entrance. The twisted frame and melted chassis of the last truck was hauled out by 2:00 a.m. on Saturday morning.

Approximately 150 people worked through the night, removing more than 650 tons of debris from the eastbound tube. Surprisingly, the road surface itself, which was assumed to be heavily damaged, proved to be in relatively good shape once the wreckage was removed. After more than five hundred feet of ceiling had been cut away and everything else that could be removed quickly had been taken out, the area was washed and painted. At 5:09 p.m. on Sunday, only about fifty-six hours after the fire began, the eastbound tube was once again open to traffic during daylight hours. The first motorists traveling through could see blackened and pockmarked walls in places and smell the lingering scent of ammonia and cleaning solutions, but otherwise the tunnel appeared to be in relatively good condition, except for the great long gap in the ceiling.

The Port Authority soon prepared notices to hand out to people traveling east through the south tube, stating that "until further notice" the tube would be open all day and night on Sundays, but from Monday through Saturday, it would be closed from 8:00 p.m. to 6:00 a.m. so that crews could go in and make repairs. During those hours, the north tube would handle two-way traffic.[27] The Port Authority quickly prepared a complete engineering construction plan for repairs and renovation, and on Tuesday, May 17, James Mitchell, Inc., the Jersey City construction contractor for the Port Authority, began replacement of the ceiling and side walls, with electrical and other contractors to assist later.[28]

Each night for the next three months, the tunnel was briefly and eerily quiet except for the whistling of air passing through the long hole in the ceiling. Within a few minutes, however, that noise would be drowned out by the roar of trucks entering from the New Jersey side, most of them pulling equipment, including a "gun" to shoot cement onto the walls, a cement mixer, two compressors, three welding machines, and several "trains" of caster-mounted pipe scaffolding. The scaffolds were tied together in groups up to 150 feet in length and positioned to face each other so that planks could be placed across them to make a platform for work on the ceiling.

The James Mitchell crews were soon joined by workers for the Cement Gun Company of Allentown, Pennsylvania; Beach Electric Company of Newark, New Jersey (electrical repair); G. M. Crocetti of Jersey City (tile replacement); and Shinick Building Company of New York (vitrified-clay duct repair). By 8:25 p.m. each night, everything would be in place, and up to ninety men would be working at full speed. The workers tended to get in each other's way and were initially bothered by the "weird whistling" of the air rushing through the huge gap in the ceiling. Several men lost their hats from the suction; but all soon managed to adjust to the various challenges of the job, and work proceeded with considerable speed. By the time the tube reopened each morning, all equipment would be gone from the tunnel, with little indication remaining of the furious effort that had taken place the previous night.[29]

In all, 536 feet of ceiling, 700 feet of side walls, 300 feet of tile duct banks with their various power and communications cables, 750 feet of walkway, and 5,000 square feet of concrete lining were completely replaced. To eliminate the need for supporting braces, which would block the travel lanes between work sessions, the engineers redesigned the ceiling so that it would be supported from above. Completed in time for resumption of full service prior to the Labor Day weekend rush of traffic, the repairs cost approximately $600,000. The cost of the accident in terms of injury and death, however, was yet to be totaled. Sixty-six people sustained injuries, most from smoke inhalation or being struck by falling concrete. Twenty-seven of the injured required hospitalization, although some of the men sent home after initial treatment may have sustained the greatest injury. Two certainly did. Chief Gunther E. Beake of the Third Battalion, FDNY, was treated for lacerations to the leg and shoulder and then dismissed. He died on August 23, 1949, from the lingering effects of smoke inhalation. Port Authority Patrolman Edward Bryan was not even mentioned in the newspapers among the list of injured, but according to his wife, Mary, he suffered for nineteen months before succumbing to injuries from the fire on December 23, 1950.

The magnitude of the event, in both its financial burden and human suffering, necessarily led to questions concerning why the fire occurred, who was to blame, and how a similar accident could be prevented in the future. Those questions began with reporters who sought to interview the driver of the truck carrying the carbon disulfide, Edward Tyndall. After being discharged from Beckman Hospital without any sign of injury, Tyndall returned to his home in Weehawken, where he refused all requests for information, saying that he was not well enough to talk. Later that afternoon, representatives from Boyce Motor Lines took him to the Plaza Hotel in Jersey City, presumably so that he could get his story straight before speaking further with the police.

On the afternoon of Saturday, May 15, Tyndall appeared at the Second Precinct of the Jersey City Police Department, accompanied by Joseph C. Gavin, a lawyer for Boyce Motor Lines. Police interrogated Tyndall for about an hour, and he gave them a three-part statement in which he basically denied knowing anything about anything. According to Abe Sepenuk, assistant Hudson County prosecutor, the process of learning the facts of the accident was complicated by Gavin, who was not cooperative.[30]

The most important statement following Tyndall's interrogation came from Acting Inspector Peter Smith, in charge of the Jersey City Police Department Detective Bureau. Smith expressed his satisfaction that Tyndall had not violated any municipal ordinance or state law and said he was free to go. Although Smith did not claim that Tyndall had *not* violated the regulations of the Port Authority or of the ICC, his statement indicated that prosecutors would point to the trucking company, and not the trucker, in their future efforts to assign blame.

Governmental officials investigating the accident, however, quickly realized that violations of Port Authority regulations were misdemeanors in both New York and New Jersey, punishable only by a fifty-dollar fine and five days in jail. At the fourteenth annual meeting of the Northwest Regional Conference on Highway and Motor Vehicle Problems, held in the Roosevelt Hotel in New York on October 20, 1949, Billings Wilson, director of operations for the Port Authority, told attendees that because existing penalties were no deterrent, unscrupulous shippers were sending prohibited chemicals through the tunnel in quantities that would "blow half of the waterfront off the map."[31] His claim was backed up by a spot check of trucks entering the tunnel, made by the Port Authority just a few days after the fire. Of approximately six thousand trucks inspected, more than one hundred had to be turned back because they violated the regulations.[32] Clearly, something needed to be done.

On Monday, May 16, Port Authority Executive Director Austin J. Tobin, acknowledging that existing penalties were "obviously inadequate," stated that he would ask the legislatures of both states to increase the severity of penalties. On June 1, two bills were introduced during a special session of the New Jersey legislature: one to increase the regulatory powers of the State Department of Labor and Industry and the other to increase maximum penalties for a violation of the Port Authority's regulations to a $2,000 fine and seven years in jail.[33] The bills were soon sidetracked, however, due to lobbying by some of New Jersey's oil and chemical companies.[34] The response from New York was even weaker, as legislators in Albany initially failed to do anything.

On July 29, two reports were released that addressed the issue of what could be done to prevent another similar accident from occurring and how to control such an event if it did occur. The first report, by New York Fire Chief Peter Loftus, who had been in the thick of the firefighting efforts on May 13, cited several inadequacies in the procedures and equipment in place to fight such a fire and also suggested that studies should be conducted to determine if the tunnel's ventilation system should operate differently in similar conditions.

The second report, by the National Board of Fire Underwriters, stated somewhat optimistically that new legal safeguards to broaden the power of the authorities "are in the process of enactment at the time of this writing" and claimed that "all indications point to more severe regulations with greater penalties for violations." But the report also emphasized, "It would be of little avail to inform the truckmen of what not to do without offering a safe alternative." Therefore, "a 'safe routing' plan for the transportation of dangerous chemicals and explosives should be established for any area in which are included vehicular tunnels, bridges, and congested highways."[35]

On October 5, 1949, a federal grand jury indicted Boyce Motor Lines on six counts of unlawfully transporting a dangerous chemical in violation of ICC regulations.[36] On January 10, 1950, the J. T. Baker Chemical Company was similarly indicted.[37] The Port Authority also sued both companies for $800,000 in damages in New York State Supreme Court. In June 1950, the Port Authority accepted $300,000 in settlement, with Boyce Motor Lines paying $50,000 of that amount.[38] Boyce Motor Lines appealed the original federal grand jury indictment, claiming that it was guilty of no more than "an honest error in judgment" and that the ICC regulation was defective in that the company's trucks could not reach their destinations in Brooklyn without traveling over or through "congested thoroughfares, places where crowds were assembled, streetcar tracks, tunnels, viaducts and dangerous crossings."[39]

On July 16, almost a year after the Loftus and National Board of Fire Underwriters reports were made public, Federal District Court Judge William F. Smith dismissed the three most important counts against Boyce Motor Lines, because the words "so far as practicable and where feasible" in the regulation were "so vague and indefinite as to make the standard of guilt conjectural."[40] One can only speculate on the impact that that decision had on Battalion Chief Gunther E. Beake's next of kin, who were presented with a posthumous award for heroism by the Uniformed Firemen's Association the following day at New York City Hall Plaza.[41] The fight, however, was not over.

The court of appeals in the Third District reversed the district court's decision, finding that the regulation established a reasonable standard of conduct. Boyce Motor Lines subsequently petitioned the U.S. Supreme

Court for a review of the case, which was decided on January 28, 1952. The majority opinion, written by Justice Tom C. Clark, held that the regulation was not too vague for proper enforcement and that the district court should not have dismissed the counts of the indictment.[42] Justice Robert H. Jackson wrote the dissenting opinion, in which he was joined by Justices Hugo L. Black and Felix Frankfurter. Jackson argued that the regulation in question did not contain a "definite standard from which one can start in the calculation of his duty," because all routes were equally open and all routes were equally closed. Essentially, Jackson's opinion recognized that transportation of explosives or inflammable liquids necessarily involved passing through many congested thoroughfares and through or over tunnels, viaducts, or bridges and that an explosion would have been equally dangerous in any of those points of constriction.[43]

The dissenting opinion may have given some comfort to Boyce, but the company still lost the case. On May 19, 1952, the district court in Newark fined Boyce $5,000 for the count involving its transportation of carbon disulfide through the Holland Tunnel on May 13, 1949, and fined the company an additional $3,000 for the transportation of dangerous chemicals on previous occasions. No jail time for individuals was imposed.[44] The fine was undoubtedly much less than what Boyce had spent defending itself and was small compared with the profits to be derived from continuing to ignore regulations. The case, however, proved to be of considerable importance in relation to establishing a standard for shipment of hazardous materials through congested areas and was cited for years to come in other cases involving similar issues.

The 1949 Holland Tunnel fire also brought into focus the inadequacy of the equipment available to the Port Authority to fight large fires and the inadequacy of its emergency procedures. Moreover, as the Loftus report recommended, the procedures for operation of the ventilation system during a fire needed to be carefully studied. Given that the Lincoln, Queens-Midtown, and Brooklyn-Battery Tunnels all used essentially the same ventilation system as the Holland Tunnel, consideration of the last recommendation was particularly urgent.

At 7:43 a.m. on November 15, 1951, just one month after the Supreme Court agreed to hear the *Boyce* case, an eastbound truck loaded with cardboard boxes containing jugs bound for the Old Dutch Mustard Company in Brooklyn caught fire about two hundred feet into the south tube of the Holland Tunnel. Dense clouds of smoke soon filled the tube, as motorists abandoned their cars and fled on foot. Due to a dock strike, traffic that morning was unusually heavy, and emergency vehicles trying to enter the tunnel from the Jersey City side were blocked by lines of cars, buses, and trucks that stretched six miles west of the entrance plaza. Manhattan traffic also snarled after the north tube opened temporarily for two-way traffic. By the time Jersey City firefighters were finally able to hook up their hoses to the tunnel's standpipes, overhead tiles, loosened by the heat, began falling down around them. Several times the firefighters thought that they had the blaze extinguished, only to see the flames burst out again. Approximately seventy-five cadets from the New York Police Academy were sent into the tube to clear away smoldering debris, and after the cleanup, it was found that some of the tile had buckled. Otherwise, the tunnel was undamaged.[45]

At about 9:00 p.m. on June 14, 1964, a tractor-trailer carrying academic caps and gowns caught fire in the south tube of the Holland Tunnel. The fire spread to the gas tank, which exploded. No one was reported injured, but the blast damaged tiles, knocked out some lights, and backed up traffic to the far end of the Pulaski Skyway for more than two hours.[46] On these occasions, there had apparently been no violation of regulations, and as was the case with the chemical fire in 1949, there was no malicious intent to cause harm. The vulnerability of New York–area vehicular tunnels to fire and explosion, however, was once again made apparent.

In 1996, Universal Pictures distributed *Daylight*, a narrative film produced by Davis Entertainment, starring action hero Sylvester Stallone. Inspired, somewhat, by the 1949 Holland Tunnel fire, the story involves events stemming from explosion of a truck carrying toxic waste through the "tunnel to New Jersey" and the subsequent fire. In this case, the tunnel roof collapses and several people die, but the character played by Stallone and a female companion are blown up through the breach, emerging, alive, in the Hudson River. Neither the Holland Tunnel nor the other tunnels serving Manhattan are as likely to suffer catastrophic collapse as the fictional tunnel of the film. They were designed and built too well. And the heroics displayed by the film's characters, though entertaining, are nothing compared to the actions of the firefighters, police officers, emergency workers, and average citizens who risked their lives in response to the fire of 1949 and to other, less serious fires that have occurred over the years. Those people take their place, along with the engineers who designed and the laborers who built these underwater links, as the real-world heroes of the vehicular tunnels of New York.

13

Built to Last Forever

It was built to last forever like the old Roman roads and it'll be good forever.

—Lawrence Lewis, Port Authority Manager of Tunnels,
November 1977, on the fiftieth anniversary of the tunnel opening

IN *EMPIRE ON THE HUDSON: Entrepreneurial Vision and Political Power at the Port of New York Authority* (2001), Jameson W. Doig refers to a recurring theme of political analysis: that government projects with "great expectations" are likely to be overwhelmed by greater obstacles, leading to disappointment or outright failure. Here, Doig is writing generally about the Port of New York Authority's bridge projects, which he deems to have been ultimately successful, and specifically about the initial failure of the Port Authority's Comprehensive Plan for improving transportation within the Port of New York. In the mid-1920s, years before groundbreaking for the George Washington Bridge, the Authority's first trans–Hudson River project, the plan still embodied the hopes of the Port Authority commissioners and engineers. But those hopes, Doig observes, "would all be dashed by decade's end."[1]

According to Doig, however, neither the failed plan nor the difficulties the Port Authority faced in fulfilling its mission undermined the regional and "nonpolitical" focus that motivated the Port Authority's commissioners and engineers and enabled them to move their projects forward. Unlike the members of the two state tunnel commissions, which Doig states "often behaved as suspicious guardians of provincial rights," the Port Authority commissioners acted with "some broader sense of the metropolitan region" and saw cooperation "as the route to economic growth." It also helped that the Port Authority could finance construction by issuing bonds. This freed it from the political uncertainties of statewide referenda or yearly legislative appropriations, allowing it "some political insulation [that was] denied the tunnel commissioners." In contrasting the Port Authority's success to the presumed failure of the tunnel commissions, Doig concludes that when the Port Authority took over the Holland Tunnel in 1930 and "swallowed" the tunnel commissions in 1931, "technical expertise was married to political strategy, to the greater glory of the bi-state agency —and perhaps to the benefit as well of economic development across the metropolitan region."[2] Perhaps. As a group, the tunnel commissioners were plagued by political imperatives throughout the planning and construction of the Holland Tunnel. Some commissioners, particularly those representing the New Jersey tunnel commission, were corrupted by very narrow political and personal interests. But to portray all the tunnel commissioners and, by extension, their engineering staff as equally ineffective in completing their task does an injustice to the majority of them and negates the honor they should be accorded for what they accomplished. They applied a high level of technical expertise and worked their way through complex political strategies years before creation of the Port Authority, and they did so to the benefit of the greater New York area.

Certainly, there were great expectations for the Holland Tunnel, and given that its completion was more than three years behind schedule and more than $20 million over budget, it probably was a disappointment to those who wanted it finished sooner and at a lower cost. But it was hardly a failure. It achieved its primary purpose: providing an all-weather alternative to the railroad-dominated ferry and barge/lighter/car-float systems for bringing motor vehicles and freight into New York City, and it did so without completely destroying those systems. That those modes of transportation did gradually decline was not due solely to completion of the Holland Tunnel, which was merely one in a multiplicity of factors leading to their eventual demise.

The Port Authority commissioners swam in the same polluted sea of politics while building their connections across the river as did the tunnel commissioners while building the first tunnel. Neither group could function without being responsive to the existing political environment. As Doig notes, the Port Authority commissioners were not "so unwise as to turn their backs entirely on the political forces linked to immediate public concerns, or the narrower traditions of political favoritism."[3] And after the Port Authority and tunnel commissions merged, the new entity included six former tunnel commissioners. Two of these, Dyer and Shamberg, had been heavily involved in planning the tunnel. Thus, the tunnel commissioners were not some entirely separate species of fish, different from their

Port Authority brethren. In addition, the newly combined staff included 480 former employees of the tunnel commissions, while having only 275 employees of the Port Authority.[4]

The merger of the tunnel commissions with the Port Authority allowed the Port Authority to use the Holland Tunnel revenues to finance its other projects. The freedom from periodic legislative appropriation that the Port Authority commissioners enjoyed would never have been possible without the millions of dollars in yearly tunnel tolls that the Port Authority collected. And whatever measure of "political insulation" they enjoyed would also never have been possible without the earlier success of the tunnel commissioners and their engineers. Doig notes this in an earlier essay. The Port Authority found itself loaded with debt from bonds issued to finance the Outerbridge Crossing, Goethals Bridge, and George Washington Bridge. It was the Holland Tunnel's toll revenue, "the largest single source of toll income from vehicular traffic in the nation," that saved the Port Authority from bankruptcy in its early years.[5] That revenue stream, which far exceeded the expectations of Holland Tunnel planners, also enabled the Port Authority to construct the Lincoln Tunnel, the second vehicular tunnel across the Hudson River, and other transportation-related projects throughout the port. Today, at least one promise made by politicians and Holland Tunnel planners to the citizens of New Jersey and New York remains unmet. Long after the tunnel's construction costs have been paid in full, it is still a tolled facility, even if tolls are now collected only from drivers heading east into New York City.

The Holland Tunnel's financial success enabled the Port Authority to reshape the physical and economic environment of the Port of New York. This may be the tunnel's most lasting legacy. Evaluating that legacy, however, is problematic. Many people opposed the Holland Tunnel upon its inception because they were convinced that it would adversely impact the regional economy. This opposition was particularly strong among those who saw the existing transportation system as being to their advantage and feared that a tunnel would favor one state to the detriment of the other. Opposition to the tunnel was as much intrastate as interstate, however. Some New Yorkers believed Manhattan would benefit far more than would the outlying boroughs in which they held commercial interests. In New Jersey, there were those who did not want to see northern counties prosper more than the state's southern region.

Constructing highways under or across the Hudson River did encourage suburbanization and commercial development in northern New Jersey.

The Holland Tunnel was just one of many transportation links, however, and other projects, such as the Manhattan Bus Terminal, also contributed to this growth.[6] Yet tax valuations for the five New Jersey counties nearest the Holland Tunnel—Bergen, Essex, Hudson, Passaic, and Union—from 1916 to 1929, before any Port Authority–constructed links across the Hudson River opened, revealed an increase of $2,235,346,257. By contrast, the total tax valuation in 1929 for all counties in the state of New Jersey was $3,556,955,365. Therefore, those five counties represented approximately two-thirds of the total tax valuations of the entire state in 1929. As Joseph G. Wright, the last chairman of the New Jersey tunnel commission, stated in April 1930, "It is not asserted that this increase is due entirely to the Holland Tunnel, but it is admitted that this connecting highway between the two States has had a greater influence upon the development of the metropolitan area located in New Jersey than any other single factor."[7]

Trans-Hudson road improvements also brought about economic change in Lower Manhattan. Here, however, trends toward the decentralization of industry and development of secondary business districts outside of "downtown" had a profound effect on economic activity and land use.[8] It is difficult, therefore, to isolate and evaluate the indirect economic effects (those that are caused by the tunnel but are later in time or farther removed in distance) and cumulative economic effects (those that result from the incremental impact of the tunnel when added to other past, present, and future actions) of the tunnel's completion.[9]

The areas surrounding the Manhattan tunnel entrance and exit did experience considerable redevelopment in the years immediately preceding and following the tunnel's completion. The three-story Tunnel Garage at the corner of Thompson and Broome Streets, one of the first garages in Manhattan constructed specifically for motor vehicles, was built to take advantage of the anticipated tunnel traffic. It even featured, on the corner of the top floor, a terra-cotta medallion showing a Model T Ford emerging from the mouth of the tunnel. But this building, as an example of redevelopment directly tied to the tunnel, is an exception. Between 1925 and 1929, millions of square feet of industrial and office space were added to the rapidly redeveloped area within a few blocks of the Holland Tunnel entrance plaza. Moreover, the occupancy rates in these buildings were among the best in Manhattan. The commonly held assumption of many people in the real estate business was that this redevelopment was primarily motivated by the tunnel project. A. L. Benel, head of the industrial development department of Brown, Wheelock, Harris, Vought & Company, which assembled and managed many of the newly developed properties in this area, disagreed. In November 1929, Benel told a *New York Times* reporter, "The real reason why the development came was that here could be found about the only block fronts in Manhattan available for such improvement with convenient shipping and rail facilities and with a subway running directly underneath."[10] Redevelopment of the area, particularly along both sides of Varick Street, actually began before firm plans for the tunnel were announced. But the availability of large parcels for development was driven by a decision made in 1925 by Trinity Church, which owned large tracts of land near the proposed tunnel entrance, to sell some of the land for speculative development. The indirect effect of the tunnel was to increase land values in the area around the portals, mainly because the associated widening of streets and construction of tunnel plazas provided more light and air to surrounding buildings.

The greatest impact of the Holland Tunnel, to both northern New Jersey and New York City, has probably been from the growth in vehicular traffic that it generated. In this regard, the tunnel was, perhaps, too successful. Conceived in the nascent years of the automobile age, the tunnel contributed to a traffic-congestion nightmare undreamt of by those who thought that the tunnel would help solve, rather than exacerbate, traffic woes. After decades of experience with operation of trans–Hudson River highways, many politicians and transportation planners had begun to realize by the 1950s that these improvements often had unforeseen adverse impacts.

In 1951, the Port Authority announced plans to build a third tube of the Lincoln Tunnel. In response to these plans, officials in Weehawken, New Jersey, and in New York City adopted a position similar to the one Jersey City Mayor Frank Hague held in the 1920s, concerning the costs of accommodating tunnel-generated traffic. They wanted the Port Authority to help pay for the extensive street and highway improvements that would be necessary to keep traffic moving. Eventually, the Port Authority agreed, and in October 1951, it approved $4 million to fund construction of connections to the New Jersey Turnpike.[11]

In February 1957, the Port Authority completed the $11 million, eight-block-long, four-lane Lincoln Tunnel Expressway, connecting the tunnel to Thirtieth Street in Manhattan. The third Lincoln tube opened to the south of the first two in May 1957, with the Port Authority hoping that the new expressway would alleviate midtown congestion.[12] It did not.

In 1956, with the third tube of the Lincoln Tunnel under construction, Robert Moses, in his new role as New York City construction coordinator, called for a third tube for the Holland Tunnel. He believed that another tube would be necessary to handle the additional traffic generated by new express highways, some of which he intended to build. The Port Authority did not agree. It assumed that a proposed Narrows Bridge linking Brooklyn with Staten Island would attract enough of the Long Island and Westchester County traffic then forced to use Manhattan streets that the existing two tubes of the Holland Tunnel would be adequate. Although Moses acknowledged the opposition, he arrogantly predicted that the new tube would be completed in four or five years.[13]

As *New York Times* reporter Joseph C. Ingraham noted, Moses and the Port Authority were missing the real issue. "Regardless of the differing viewpoints over the Holland Tunnel," Ingraham wrote, "the question remains: Can the constant demand for new roads to accommodate the inexorable rise in the annual output of cars ever be satisfied?"[14] The question applied equally to the relatively short segments of road running through tunnels as it did to cross-town expressways. The chief design engineer of the Holland Tunnel, toward the end of his career, came to the conclusion that as far as tunnels were concerned, the answer to Ingraham's question was no. Although Ole Singstad remained proud of his tunnel design work throughout his life, he eventually came to doubt the efficacy of vehicular tunnels as a means of addressing New York City's expanding transportation needs. In July 1967, a reporter for the *New York Times* interviewed Singstad, then eighty-five, in the elderly engineer's small corner office overlooking the Battery, from which Singstad had a good view of the New York ventilation towers of the Holland and Brooklyn-Battery tunnels "and a dim view of the thousands of cars that stream out of the tunnels every day." Looking back at decisions made decades before, Singstad said, "I think we've overdone it. The city is choking itself to death with autos." Without directly addressing the freight-transportation problems that created the primary need for New York's first vehicular tunnel but focusing more on the (once) secondary purpose of vehicular tunnels as a means of bringing people in and out of the city (via private cars), Singstad suggested, "The city should build more subways. Do you realize that we have had no subway construction to mention since the nineteen-thirties? We should build more rail tunnels under both the Hudson and East Rivers."[15]

The greatest failure of the engineers who planned the Holland Tunnel was to underestimate the volume of traffic it would eventually carry and the extent to which passenger vehicles would supplant trucks as the dominant motor vehicle using the facility. When Clifford Holland made his first report as chief engineer in December 1919, it was assumed that a tunnel of the design he recommended, accommodating only motor vehicles (no horse-drawn wagons) in both directions, could handle 15.8 million vehicles a year at full capacity (which was projected to be reached in its fifteenth year of operation).[16] When the tunnel opened in 1927, the estimated capacity had been reduced to 15 million vehicles annually, partially in expectation that other tunnels would be opened before it reached its original full-design capacity, thus relieving it of the burden of accommodating all potential demand. In 1957, its thirtieth year of operation, it was already carrying more than 20 million vehicles a year.[17] It currently carries more than 34 million vehicles a year. The total eastbound traffic in 2009 was slightly more than 16.6 million, and the total annual traffic in both directions during the past five years has ranged from slightly less than 34 million to about 34.7 million. This is less than the annual traffic carried by the other vehicular links between northern New Jersey and Manhattan, the Lincoln Tunnel (nearly 42 million) and the George Washington Bridge (106 million), but much more than even the most farsighted engineers had anticipated in 1919.

Historian John B. Rae has found that mass ownership of motor vehicles led directly to the boom in American road construction in the twentieth century, resulting in great achievements in civil engineering, of which the Holland Tunnel is one example. In recent years, however, scholars have challenged the work of Rae and others who see motor vehicles as deterministic in development of American road-building policy. In so doing, they have also refuted the Progressive-era notion that engineers as policymakers were objective and apolitical in their decision-making. Yet, as leading proponent of the engineer-as-politician school of thought Bruce E. Seely admits in *Building the American Highway System: Engineers as Policy Makers* (1987), "Many technical experts sincerely believe that they can make objective decisions for the benefit of society, and on occasion they have been able to live up to this claim. Thus, for many reasons the idea [of apolitical technical expertise] gained a life of its own after the Progressive period."[18]

According to Seely, the Progressive period's zeal for reform began to collapse in the aftermath of World War I. Although the detailed planning for the Holland Tunnel began during the waning of the Progressive period, the tunnel was clearly a testament to the belief that objective application of technical knowledge could triumph over politics. Whether the traffic-generating tunnel is ultimately a benefit or a bane to the society it serves is almost irrelevant in assessing the work of the people who designed and built it. Completion of the facility is, by any measure, a landmark event in the long history of automobile-related infrastructure creation, and the tunnel is one of the most notable engineering products of the Progressive era.

The product remains, but the people who brought the Holland Tunnel into being are long gone. Only a few of them are memorialized or even remembered. The first chief engineer is known because the tunnel is named for him. A bronze bust of Clifford Holland was dedicated at the New York entrance of the westbound tube in November 1953. The New York entrance plaza was named Freeman Square in honor of Milton Freeman, the second chief engineer, even before the tunnel opened. Changes over the years have made the plaza less of a fitting memorial, however, and it is doubtful that many motorists are aware of the honor.

The tunnel's design engineer and third chief engineer, Ole Singstad, received many honors throughout his long career, although they were as much for his cumulative contributions to the art and science of tunnel engineering as they were for his work on the Holland Tunnel. Singstad worked as a consulting engineer on dozens of tunnel projects around the world, including the George A. Posey Tube under the Oakland-Alameda Estuary (1928) (the first vehicular tunnel built by the trench method) and the Baltimore Harbor Tunnel (1957) (a twin-tube design, also built by the trench method, that eliminated one of the nation's worst traffic bottlenecks). In addition to his design work, he taught as a visiting lecturer at Harvard University and at New York University and served as president of the American Institute of Consulting Engineers from 1941 to 1943. He also made a significant impact on the practices of his profession by writing the "Canon of Ethics for Engineers." He died in New York, the city served by the greatest number of his projects, on December 8, 1969.[19]

Aside from Holland and Singstad, William Wilgus, chairman of the Board of Consulting Engineers from 1919 to 1922, was the most influential engineer in development of initial plans for the tunnel. In 1924, Wilgus accepted a position as consultant to the Committee on the Regional Plan of New York and Its Environs. In that capacity, he attempted to convince city planners and politicians throughout the 1920s and 1930s that radical measures were necessary to solve the transportation problems of the city. His plans, however, were too radical to win widespread support and never came to fruition.[20] When he died in 1949, his obituary in the *New York Times* noted that Wilgus was a man of many ideas and many causes, and though successful in most, he did not forget those he counted "lost," which included his plans for a small-car freight subway in Manhattan.[21]

The names of the bridge and tunnel commissioners are largely forgotten. When former New York Bridge and Tunnel Commission Chairman George Dyer died in 1934, while serving as chairman of the Port of New York Authority, there was talk of renaming the Midtown Hudson River tunnel for him. Instead, he was accorded a more modest honor. There are probably few New Yorkers who know why the little street between Thirty-Fourth Street and Forty-Second Street that helps funnel traffic to the Lincoln Tunnel is named Dyer Avenue or what its namesake contributed to the creation of the Holland and Lincoln Tunnels.

When New Jersey Interstate Bridge and Tunnel Commission member John Boyle died in 1930, Dyer served as an honorary pallbearer. Apparently, he was not the kind of man who held a grudge. But there is no telling what he would have thought about the renaming, in 1936, of Fourteenth Street in Jersey City between the exit of the Holland Tunnel and Jersey Avenue as Boyle Plaza. A memorial printed at the time lauds Boyle for having saved Jersey City taxpayers thousands of dollars by insisting that the Jersey City tunnel plazas be paid for by the tunnel commissions. Nothing was said about the hundreds of thousands of dollars he cost the taxpayers of both states by instigating or supporting needless delays. In the 1920s, Boyle was probably best known in New Jersey as owner of "Boyle's Thirty Acres," the location of a fight between heavyweight champion Jack Dempsey and Georges Carpenter in 1921. Now the exit plaza to the Holland Tunnel, the site of a battle fought with words instead of fists, serves as his memorial.

Boyle's fellow commissioner and partner in mischief, T. Albeus Adams, attended the Boyle Plaza dedication ceremony. Adams probably did more than any other commissioner to bring about passage of the tunnel-enabling legislation in both states, albeit with his own benefit in mind. But his efforts to influence the course of events for other purposes should also be noted. After the tunnel was built, from 1930 until his death in 1940, Adams served as president of the Marketmen's Association of the Port of New York, and in 1932 he helped organize the Warehousemen's Protective Committee, which he subsequently served for seven years as chairman. This group played a prominent role in fighting the invasion of railroad corporations into the commercial warehouse business and resisting the reduction of warehouse rates to levels below costs of service.[22] To whatever extent he is remembered by history, it is probably due to this and associated activity and not to what he did, meritorious or shameful, in regard to the tunnel.

A. J. Shamberg was the leading force in creation of the first Hudson

River bridge and tunnel commissions in 1906. He later served as a commissioner of the Port of New York Authority, even though he opposed placing the tunnel under Port Authority control.[23] His association with the Holland Tunnel spanned thirty-five years and exceeded that of any other commissioner. He is not well-known today, however, and neither are the other bridge and tunnel commissioners.

Walter Edge felt that his achievement in advancing the tunnel project ranked among the most notable of his political career. Yet, like the other politicians who had a hand in bringing the tunnel about, his efforts are overshadowed by the totality of his life's work. After his success in promoting enabling legislation in Trenton and Albany, he resigned as governor of New Jersey in 1919 to represent his state in the U.S. Senate, where he served until 1929. In 1929, he accepted appointment as U.S. ambassador to France and retained that position until 1933, when he returned to the world of business. He was elected to a second three-year term as governor of New Jersey in 1944 and, in that capacity, played a key role in the eventual demise of Jersey City boss Frank Hague. Edge appointed Walter D. Van Riper as attorney general of New Jersey in 1944. Van Riper seized control of the Hudson County prosecutor's office, which led to the indictment of many Hague appointees. Another blow to Hague's power came in 1947 when he lost control over judicial appointments in Hudson County.[24] Hague finally resigned in June 1947.

As for the lower-level engineers, contractors, clerks, sandhogs, and common laborers who worked on the project, their contributions, for the most part, will never be individually acknowledged. The names that we do have of those who died building the tunnel may be found in the pages of this book.

What these men built endures, although not exactly as it was originally constructed. As the Holland Tunnel has aged, essential parts of it have been refurbished, replaced, or upgraded. There have also been changes to associated buildings and street systems. The first major alteration took place less than a year after it opened, with installation of an "electronic brain" designed by Westinghouse Electric and Manufacturing Company. This system, costing about $100,000, simplified the job of monitoring and controlling traffic signals, lights, pumps, ventilation fans, and other equipment, allowing two people to do the work formerly handled by six.[25]

Numerous changes were made in the 1950s, including replacing the original external lighting fixtures; constructing a new service building and renovating the administration building; replacing the granite roadway surface with asphalt; and constructing a new $2.5 million traffic-circulation rotary for the New York exit in 1958. In August 1970, one-way tolls were instituted on twelve Hudson River and New York–New Jersey bridges and tunnels, including the Holland Tunnel. After the new collection system was deemed a success, the New York toll booths of the tunnel were removed the following year. The Port Authority also completed an extensive $500,000 electrical-system modernization early in 1972.

In 1987, the Port Authority completed a four-year, $78.3 million replacement of the original cast-concrete ceiling with nearly four thousand precast, steel-reinforced concrete panels, each measuring roughly twenty feet by five feet. The panels were covered on one side with ceramic tile manufactured in Spain. The following year, the Port Authority replaced the eight toll booths in Jersey City with nine completely new booths at a cost of about $54 million. Contractors also replaced the bronze metal finishing around utility openings inside the tunnel with stainless steel and modified the curb drains. The electrical system was also upgraded in the 1980s, and all the original incandescent interior light boxes were replaced with continuous lines of fluorescent lighting. In 1992, the Port Authority completed an extensive renovation of the New Jersey toll plaza and administration building.

Major physical changes to the tunnel over the past decade include modernization of the fire-protection, motorist-information, and ventilation systems. But the most significant changes to the tunnel since it opened have been operational rather than structural. Some of these changes have been experimental, such as the use of miniature electric-powered patrol cars on the sidewalk in 1954, while others, like the abolishment of toll collection at the New York entrance, are likely to be permanent. The most profound operational changes, however, relate to fundamental conceptual shifts regarding how the tunnel should function.

Following terrorist attacks on the World Trade Center in September 2001, both the Holland and Lincoln Tunnels were closed to all but emergency traffic. The Lincoln Tunnel soon reopened for traffic going in both directions, but it was not until October 15, 2001, that traffic inbound to Manhattan through the Holland Tunnel once again flowed through, and then only with severe restrictions on single-occupant and commercial vehicles. The Port Authority lifted the single-occupant restriction in November 2003, but following threats of terrorism in August 2004, all trucks were prohibited from entering the eastbound tube to New York. At present, all commercial vehicles are prohibited from entering Manhattan through the tunnel, but two- and three-axle, single-unit trucks may travel westbound to New Jersey at all times. The current restrictions are primarily part of a plan to manage traffic congestion in Lower Manhattan, but they also provide benefits in safety. Shippers are now much less likely to send potentially hazardous cargos through the tunnel and have adjusted their routes to use other transportation links. The tunnel is also, no doubt, safer from the potential actions of terrorists.

Perceptions that the Holland Tunnel could be vulnerable to attacks or used in some way by those with evil intent are nothing new. In 1935, police apprehended two Japanese tourists for taking photographs of the tunnel. In 1950, a motorist driving down the West Side Highway called the Port Authority to report, "There is a strange-looking Chinese vessel lying over the Holland Tunnel, maybe getting ready to do something." The call was relayed to federal officials, who raced to the waterfront but did not find a ship over the tunnel. Traveling uptown to the Lincoln Tunnel, they saw the source of the motorist's concern. A United States Lines freighter, although flying an American flag, also displayed some Chinese characters as part of the company's effort to promote its business in Far East ports.[26]

A more serious bomb threat closed the westbound Holland tube for twenty-three minutes during the evening rush hour on July 17, 1974, and there have been other threats over the years that never came to the public's attention. In 2006, there were news-media reports of a plot to blow up the tunnel with the intention of flooding the Manhattan financial district. Further investigation revealed that the scheme was actually aimed at the PATH tunnels and not the Holland Tunnel.[27]

Current police vigilance and awareness of security concerns prevents the sort of idle gawking at the tunnel's entrances and exits that was common in the months after it opened. Anyone who stands outside the tunnel and stares these days, even for the innocent purpose of gathering information for a book, may be questioned by police, as I was.

Whether due to a need to manage traffic congestion or a desire to lessen security concerns, the prohibition of commercial traffic in the eastbound tube of the Holland Tunnel may never be lifted. If that is the case, one tube of the tunnel will never again serve the primary purpose for which it was built. The facility that the New York Times referred to in 1917, when it was still in the planning stages, as the "Hudson Truck Tube" will serve, primarily, passenger vehicles.[28]

Some dreams die hard. Plans for a better system of freight distribution for the Port of New York are at least as old as William Wilgus's report to the

Amsterdam Corporation in 1908 and as current as the present day. Some relatively recent schemes involve rail tunnels, and some do not. But as Port Authority Manager of Tunnels Lawrence Lewis opined on the eve of the Holland Tunnel's fiftieth anniversary in 1977, it is highly unlikely that another vehicular tunnel will ever be built into Manhattan due to the prohibitive cost and the difficulty of obtaining environmental and social approval. Moreover, as Lewis said, "There's no place to put the traffic in Manhattan now. Those days are past when you can provide a tunnel for one man in a car."[29]

Lewis was mistaken in assuming that the tunnel was provided for "one man in a car," but he was correct in thinking, even as long ago as 1977, that modern concerns would probably prevent construction of another vehicular tunnel into Manhattan. Yet the Holland Tunnel will continue to serve as an important link in the transportation system of the greater New York area as long as it exists. The great majority of future motorists traveling through the Holland Tunnel will probably take its existence for granted, just as most people do in the present day. They should not. It behooves us as a society to reconsider, from time to time, the debt we owe to those who came before us for what they have left behind. To remind us of that debt, perhaps there should be a sign at each entrance to the tunnel repeating the words uttered by Walter Edge on the day of opening in 1927: "This is probably one of the greatest engineering feats in the world," he said. "You have built a very good tunnel. Now use it."[30]

Acknowledgments

I wish to thank the staff at NYU Press, including Steve D. Maikowski, director; Margie Guerra, assistant to the director and subsidiary rights administrator; and Eric Zinner, assistant director and editor in chief, for their assistance and understanding during the complex process of manuscript development. Andrew Katz, copyeditor, caught many of my errors and substantially improved the manuscript. Jeffrey L. Meikle, professor of American studies, University of Texas, recommended me to NYU Press as a potential author of a history of the Holland Tunnel. I thank him for his confidence in me and for offering advice at critical points in writing the book. The staffs of the New York State Library in Albany, New York; the Jersey City Free Library in Jersey City, New Jersey; and the New York Public Library in New York City were invaluable to me in my research. I especially appreciate the help of reference archivist Thomas Lannon, Manuscripts and Archives Division, New York Public Library. Thomas devoted many hours to scanning newspaper articles for me and greatly facilitated my research. In Cleveland, Ohio, Eleanor Blackman, special collections archivist, and N. Sue Hanson, head of special collections, Kelvin Smith Library, Case Western Reserve University, were most helpful in providing access to the Holland Collection. Dario A. Gasparini, Ph.D., P.E., professor of civil engineering, Case Western Reserve University, provided invaluable engineering insight into some of the tunneling techniques used in construction of the tunnel. I will always be indebted to Dario and his wife, Linda, for their hospitality during my research visit to Cleveland. Most of all, my thanks go to Ysabel de la Rosa for her expert editorial commentary on a first draft of the manuscript. I can't imagine having written this book without her unflagging encouragement and support.

Notes

Notes to the Introduction

1. U.S. Department of Transportation, Federal Highway Administration, "Highway Statistics Summary to 1995," http://www.fhwa.dot.gov/ohim/summary 95/mv200.pdf. The numbers presented in this report are estimates drawn from state authorities and other sources and may differ from those found in histories of the automobile in the United States, statistics that may have been taken from reports by automobile associations or other contemporary sources.

2. These buses were actually gasoline-electric hybrids. See "Motors May Replace Fifth Avenue Stages," *New York Times*, September 16, 1905.

3. U.S. Department of Transportation, Federal Highway Administration, "Highway Statistics Summary to 1995."

4. "Rival Auto Shows Vie for Patronage," *New York Times*, January 14, 1906; "Enormous Auto Sales Made at Both Shows," *New York Times*, January 21, 1906.

5. James J. Flink, *The Car Culture* (Cambridge: MIT Press, 1975), 52.

6. "New York Autoists Lead," *New York Times*, September 5, 1907.

7. Flink, *Car Culture*, 53.

8. Christopher Finch, *Highways to Heaven: The Auto Biography of America* (New York: HarperCollins, 1992), 69–72.

9. "Car Tendencies for New Models," *New York Times*, January 1, 1912; "Motor Trucks on Display at Garden," *New York Times*, January 15, 1912; "Motor Truck Show Ends Auto Season," *New York Times*, January 21, 1912.

10. "Two More Days for Motor Truck Show," *New York Times*, January 19, 1912.

11. "For Women Motorists," *New York Times*, February 22, 1912.

12. "500 Motor Trucks in Annual Parade," *New York Times*, April 14, 1912.

13. For more information on the state of automobile ownership and production in 1917 and 1918, see "Car for Every 29 Persons," *New York Times*, March 18, 1917; "Best Motor Year for Auto Makers," *New York Times*, January 6, 1918; "New York's Automobile Growth," *New York Times*, March 3, 1918.

14. New York, New Jersey Port and Harbor Development Commission, *Joint Report with Comprehensive Plan and Recommendations* (Albany, NY: J. B. Lyon, 1920), 5, 118, 140; U.S. Congress, Senate, Committee on Interstate Commerce, *Tunnel under Hudson River*, 65th Cong., 3d sess., December 12, 1918, 14.

15. John B. Rae, *The Road and the Car in American Life* (Cambridge: MIT Press, 1971), 102.

Notes to Chapter 1

1. Walter Evans Edge, *A Jerseyman's Journal: Fifty Years of American Business and Politics* (Princeton: Princeton University Press, 1948), 89–90; Joseph Bucklin Bishop, *Goethals: Genius of the Panama Canal; A Biography* (New York: Harper and Brothers, 1930), 402–404.

2. *Dictionary of American Biography*, s.v. "Goethals, George Washington"; "Memoir of George Washington Goethals," *Transactions of the American Society of Civil Engineers* 93 (1929): 1813–1817; Bishop, *Goethals*, 403.

3. Bishop, *Goethals*, 271.

4. David McCullough, *The Path between the Seas: The Creation of the Panama Canal, 1870–1914* (New York: Touchstone, 1977), 536.

5. "Memoir of George Washington Goethals," 1816.

6. Edge's other opponent in the Republican primary was George L. Record, a noted progressive. Had Record not been in the race, the twenty-nine thousand votes cast for him might have gone to Colgate, and Edge would have lost the primary. See Joseph H. Mahoney, "Walter Evans Edge," in *The Governors of New Jersey, 1664–1974: Biographical Essays*, ed. Paul A. Stellhorn and Michael J. Birkner (Trenton, NJ: New Jersey Historical Commission, 1982), 187.

7. Nelson Johnson, *Boardwalk Empire: The Birth, High Times, and Corruption of Atlantic City* (Medford, NJ: Plexus, 2002), 85.

8. Dayton David McKean, *The Boss: The Hague Machine in Action* (Boston: Houghton Mifflin, 1940), 66, 69. McKean claims that Democrats were not urged to vote Republican but were simply urged not to vote at all.

9. Edge, *Jerseyman's Journal*, 90; see also Bishop, *Goethals*, 402, 404.

10. Ibid.

11. Edge, *Jerseyman's Journal*, 93; "New Jersey Bills for Hudson Tunnel," *New York Times*, July 21, 1918.

12. Charles H. Winfield, *History of the County of Hudson, New Jersey, from Its Earliest Settlement to the Present Time* (New York: Kennard and Hay, 1874), 232–234.

13. "New Jersey Ferries," *New York Times*, August 7, 1870.

14. Edmund Walter Miller, *From Canoe to Tunnel: A Sketch of the History of Transportation between Jersey City and New York, 1661–1909: A Souvenir of Tunnel Day, July 19, 1909* (Jersey City, NJ: A. J. Doan, 1909), 4.

15. New York, New Jersey Port and Harbor Development Commission, *Joint Report*, 93.

16. Winfield, *History of the County of Hudson, New Jersey*, 252.

17. Miller, *From Canoe to Tunnel*, 7.

18. "Blockaded by Heavy Ice," *New York Times*, January 12, 1893.

19. New York, New Jersey Port and Harbor Development Commission, *Joint Report*, 106, 107.

20. Ibid., 107.

21. Ibid., 107, 138.

22. Ibid., 112, 113, 135, 136. The site of the old terminal building became the rotary of the Holland Tunnel exit.

23. Ibid., 130, 287; "Why New York Needs a Vehicular Tunnel," *New York Times*, March 9, 1919.

24. New York, New Jersey Port and Harbor Development Commission, *Joint Report*, 135. There were also freight terminals in Brooklyn and the Bronx, but these were of limited capacity, and much of the freight they handled was trucked in from terminals in Manhattan.

25. "Why New York Needs a Vehicular Tunnel."

26. Anthony J. Bianculli, *Trains and Technology: The American Railroad in the Nineteenth Century*, vol. 4, *Bridges and Tunnels, Signals* (Newark: University of Delaware Press, 2003), 108. Bianculli cites 1808 as the date of the earliest proposal for a railroad tunnel under the river.

27. "Hudson River Tunnel," *New York Times*, July 27, 1876.

28. "The Hudson River Tunnel," *New York Times*, March 2, 1875.

29. "The First Hudson Tunneler," *New York Times*, February 28, 1908.

30. "Ferry Rush Checked by Hudson Tunnel," *New York Times*, February 29, 1908; see also "McAdoo Tunnel Ready on Tuesday," *New York Times*, February 23, 1908; "100,000 Take Trip under the Hudson," *New York Times*, February 27, 1908.

31. Several bridges crossed the Harlem River into Manhattan, but they had a negligible effect on the transportation of people and freight into Lower Manhattan.

32. New York, New Jersey Port and Harbor Development Commission, *Joint Report*, 323.

33. Steven Hart, *The Last Three Miles: Politics, Murder, and the Construction of America's First Superhighway* (New York: New York Press, 2007), 20.

Notes to Chapter 2

1. New York State Bridge and Tunnel Commission, *Eighth Report of the New York State Bridge and Tunnel Commission to the Legislature of the State of New York, 1918* (Albany, NY: J. B. Lyon, 1918), 3 (hereafter *Eighth Report of the New York Commission, 1918*); New York State Bridge and Tunnel Commission, *Report of the New York State Bridge and Tunnel Commission to the Governor and Legislature of the State of New York, 1920* (Albany, NY: J. B. Lyon, 1920), 3 (hereafter *Report of the New York Commission, 1920*). Two private bridge companies existed at this time. The New York and New Jersey Bridge Company (of New Jersey) received a charter

from New Jersey in 1868, and the New York and New Jersey Bridge Company (of New York) received a charter from New York in 1890. These two companies consolidated in 1892 and received federal authorization to build a bridge in 1894. The New York charter expired in 1907. The North River Bridge Company (formed 1887) was chartered by an act of Congress in 1890, but it never received a charter from either New York or New Jersey. Its congressional charter expired on January 1, 1912.

2. "Brief History of the New York State Bridge and Tunnel Commission from inception to date," n.d., photocopy of typewritten manuscript, box 6, folder 2, C. M. Holland Collection, Special Collections, Kelvin Smith Library, Case Western Reserve University. Used with permission.

3. "Plans for Hudson Bridge," *New York Times*, March 19, 1914.

4. "Tunnels Not Bridge Favored to Jersey," *New York Times*, April 22, 1913.

5. "Decides on 57th St. for Hudson Bridge," *New York Times*, March 20, 1913.

6. "Tunnels Not Bridge Favored to Jersey"; "Tunnel Highways to Jersey Likely," *New York Times*, November 18, 1913.

7. "Comprehensive Terminal Plan Suggested Solution of Manhattan Freight Congestion," *New York Times*, February 25, 1917; "Outlook for North River Links to New Jersey," *New York Times*, June 17, 1917.

8. "Outlook for North River Links to New Jersey."

9. *Report of the New York Commission, 1920*, 56, 57.

10. "Tunnels Not Bridge Favored to Jersey"; *Eighth Report of the New York Commission, 1918*, 6, 7.

11. *Eighth Report of the New York Commission, 1918*, 7.

12. Public Service Corporation of New Jersey, *Report to the Executive Committee of the Public Service Corporation of New Jersey on the Proposed Vehicular Tunnel between the Cities of Jersey City and New York by the Special Committee Appointed to Investigate the Subject*, March 1917, 7–11 (hereafter *Report to the Executive Committee of the Public Service Corporation, 1917*); "New Board to Plan for Jersey Tunnels," *New York Times*, June 9, 1917; *Eighth Report of the New York Commission, 1918*, 4, 5.

13. *A History of the City of Newark New Jersey, Embracing Practically Two and a Half Centuries, 1666–1913*, vol. 3 (New York: Lewis, 1913), 589; New York, New Jersey Port and Harbor Development Commission, *Joint Report*, 328.

14. "T. N. McCarter Sr., Founded Utility," *New York Times*, October 24, 1955.

15. Ibid.

16. *Report to the Executive Committee of the Public Service Corporation, 1917*, 83.

17. Ibid., 88.

18. Edge, *Jerseyman's Journal*, 93, 94.

19. "Hudson River Tunnel Plan," *New York Times*, April 8, 1917.

20. Bishop, *Goethals*, 408.

21. "Would Equip Port for Trade of the World," *New York Times*, March 2, 1917.

22. Ibid.

23. "Advocates Hudson River Tunnel," *New York Times*, March 11, 1917; "Urges Co-operation in State Trade," *New York Times*, April 15, 1917.

24. "Oppose Any Closer Link with Jersey," *New York Times*, March 13, 1914.

25. "New Jersey Bills for Hudson Tunnel," *New York Times*, July 21, 1918.

26. Bishop, *Goethals*, 405.

27. "New Board to Plan for Jersey Tunnels," *New York Times*, June 9, 1917.

28. Ibid.

29. *Eighth Report of the New York Commission, 1918*, 20–30.

30. See "Memorandum of Statements made by Mr. John F. O'Rourke in conference with Colonel Wilgus," n.d., box 49, William J. Wilgus Papers, Manuscripts and Archives Division, New York Public Library, Astor, Lenox and Tilden Foundations (hereafter Wilgus Papers); used with permission. In *Goethals, Genius of the Panama Canal*, Bishop claims that it was only after Goethals had conceived the idea of a concrete-block tunnel that he discovered O'Rourke was already working on such a plan. Goethals, for his part, sent a memorandum to the joint tunnel commissions, dated April 2, 1920, in which he stated that he performed a preliminary study of tunnel costs, using precast concrete blocks in place of cast iron for the tunnel ring, and found that great cost savings were possible. "At this point I took the matter up with the head of a firm of contracting engineers who had done more actual work in the North and East Rivers than any other concern, covering years of experience. I found him an advocate of the block tunnel for the North River." (See *Engineering News-Record* 84, no. 15 [April 8, 1920]: 729). Goethals's repeated references to O'Rourke as a contractor, and not a design engineer, indicate an attempt to bolster his own claims as designer and lessen the role played by O'Rourke. He stated in testimony before the United States Senate Committee on Interstate Commerce, on December 12, 1918, that the estimated cost of the tunnel was determined by a bid. "A bid was submitted by a prominent tunnel contractor in New York on the general scheme, and they guarantee to complete it for $12,000,000." See U.S. Congress, Senate Committee on Interstate Commerce, *Tunnel Under Hudson River: Hearing on S. 4765*, 65th Cong., 3d sess. (1918), 1–26. This statement indicates that Goethals had only a general plan for the tunnel until O'Rourke presented a specific design on which he could guarantee a set cost of construction.

31. "John F. O'Rourke, Engineer, Is Dead," *New York Times*, July 30, 1934; "Associates Mourn at O'Rourke Rites," *New York Times*, August 2, 1934. Unless otherwise cited, information about O'Rourke is taken from various articles in the *New York Times*, from patent applications filed with the United States Patent Office, and from documents in the Wilgus Papers. See also the collection of photographs of the O'Rourke Engineering & Construction Company, Heller Transportation Collection, Long Island University, Brooklyn Campus.

32. Halbert P. Gillette, *Handbook of Cost Data for Contractors and Engineers*, 2nd ed. (New York: McGraw-Hill, 1910; revised 1918), 1577–1578.

33. Reprinted in *Compressed Air Magazine* 23, no. 5 (May 1918): 8739–8743.

34. "Asks Roadbuilding as Military Asset," *New York Times*, December 16, 1917.

35. New York, New Jersey Port and Harbor Development Commission, *Joint Report*, 55.

Notes to Chapter 3

1. "Coal Crisis Acute; Cold Near Zero," *New York Times*, December 29, 1917.

2. New York, New Jersey Port and Harbor Development Commission, *Joint Report*, 403; "Coal Crisis Acute; Cold Near Zero."

3. "Riot When Supply of Coal Gives Out," *New York Times*, December 13, 1917.

4. "Six below Zero Finds City Coal at Famine Stage," *New York Times*, December 30, 1917.

5. "Use of Tubes Gives City Relief at Once," *New York Times*, January 2, 1918.

6. "New York's Coal in Jersey," *New York Times*, January 15, 1918.

7. "City Coal Supply Now Up to Normal," *New York Times*, January 5, 1918; "Why Coal Shortage Hit New York Worst," *New York Times*, January 6, 1918.

8. "Six below Zero Finds City Coal at Famine Stage"; "Mercury 6 Below: Likely to Go Lower," *New York Times*, December 30, 1917.

9. "Six below Zero Finds City Coal at Famine Stage."

10. "Coal Crisis Acute; Cold Near Zero."

11. "Six below Zero Finds City Coal at Famine Stage."

12. Ibid.

13. "River Still Holds Coal from City," *New York Times*, January 16, 1918.

14. "13 below Zero on Coldest Day; Worst Is Over," *New York Times*, December 31, 1917. The Weather Bureau was formed in 1870.

15. "Urges Use of Tunnels," *New York Times*, January 1, 1918.

16. Ibid.

17. "Six below Zero Finds City Coal at Famine Stage."

18. "Relief Seekers Few with City Shivering," *New York Times*, December 31, 1917.

19. "Temperature Rising, but Cold Still Keen," *New York Times*, January 3, 1918.

20. "New York's Coal in Jersey"; "Shift Blame for Shortage of Coal," *New York Times*, January 15, 1918.

21. "Six below Zero Finds City Coal at Famine Stage."

22. "Urges Use of Tunnels."

23. "Use of Tubes Gives City Relief at Once."

24. Although the Pennsylvania Railroad stated, while the tunnel franchise was under consideration by the New York City board of aldermen (city council), that it had no need, or interest, in bringing freight through the tunnels, a clause was

inserted into the original text of the enabling document that would have allowed such use. This clause was opposed by several aldermen who feared that allowing freight movement through the tunnels would harm the economy of the Port. The clause was therefore deleted prior to passage of the franchise on December 16, 1902. See "Pennsylvania Tunnel Franchise Opposed," *New York Times*, November 27, 1902; "Letter from Mr. Cassatt," *New York Times*, December 9, 1902; "Pennsylvania Tunnel Franchise Reported," *New York Times*, December 9, 1902; Carl W. Condit, *The Port of New York: A History of the Rail and Terminal System from the Beginnings to Pennsylvania Station* (Chicago: University of Chicago Press, 1980), 269–273.

25. "Attribute 12 Deaths to Extreme Weather," *New York Times*, January 1, 1918.

26. "Use of Tubes Gives City Relief at Once."

27. "First Coal Arrives by Pennsylvania Tunnel," *New York Times*, January 3, 1918.

28. "Drastic Action Increases City Supply of Coal," *New York Times*, January 4, 1918.

29. "City Coal Supply Now Up to Normal."

30. "Drastic Action Increases City Supply of Coal."

31. Ibid.

32. "Shift Blame for Shortage of Coal"; "New York's Coal in Jersey."

33. "New York's Coal in Jersey."

34. "City Coal Supply Now Up to Normal."

35. "Shift Blame for Shortage of Coal."

36. "Favor Hudson Tubes for West Side Plan," *New York Times*, January 20, 1918.

37. Ibid.

38. "Tubes under River in Terminal Plan," *New York Times*, February 1, 1918.

39. "The West Side Report," *New York Times*, February 2, 1918.

40. "Tubes under River in Terminal Plan"; "Citizens Union Fights Plans for West Side," *New York Times*, February 26, 1918.

41. "Major Gen. Goethals Favors Hudson River Tunnel," *New York Times*, January 27, 1918.

42. "Favor Hudson Tubes for West Side Plan."

43. Erwin W. Bard, *The Port of New York Authority* (New York: Columbia University Press, 1942), 26; *New York Times*, "Port Commission Reports," February 19, 1918.

44. New York, New Jersey Port and Harbor Development Commission, *Joint Report*, 59.

45. Ibid., 329–330.

46. Ibid., 243.

47. *Eighth Report of the New York Commission, 1918*, 9.

48. Ibid., 14.

49. Ibid., 17.

50. "Work Jointly for Tunnel," *New York Times*, June 15, 1918.

51. "Explains Tunnel Bill," *New York Times*, June 28, 1918; "Ask Nation to Share in Tunnel to New Jersey," *New York Times*, June 29, 1918.

52. "Denies Fare Rise on Jersey Trolleys," *New York Times*, July 13, 1928; "Edge Replies to McCarter," *New York Times*, July 21, 1918; "McCarter Makes Apology," *New York Times*, July 25, 1918.

53. "Realty Deals Involved in Bitter Fight over Hudson Vehicular Tube," *New York Evening Post*, May 25, 1921.

54. "T. A. Adams Succeeds McCarter," *New York Times*, August 12, 1918.

55. Unless otherwise cited, information on Adams's early life comes from the following: *Scannell's New Jersey's First Citizens: Biographies and Portraits of the Notable Living Men and Women of New Jersey with Informing Glimpses into the State's History and Affairs*, vol. 1 (Paterson, NJ: J. J. Scannell, 1917), 4–6; "T. A. Adams Dead; Retired Executive," *New York Times*, September 15, 1940.

56. "C. W. Morse Buys a Bank," *New York Times*, March 27, 1905; Jay Shockley, *Gansevoort Market Historic District Designation Report*, ed. Mary Beth Betts (New York: New York City Landmarks Preservation Commission, 2003), 12. Adams was president of the bank from 1898 but did not have a controlling interest until 1901.

57. "The Triple Bank Merger," *New York Times*, April 22, 1903.

58. Thomas Hamilton Murray, *Booklet of Information Regarding the American-Irish Historical Society* (Boston and New York: American-Irish Historical Society, 1905) 17; Shockley, *Gansevoort Market*, 12.

59. "Realty Deals Involved in Bitter Fight over Hudson Vehicular Tube."

60. Ibid.

Notes to Chapter 4

1. Senate Committee, *Tunnel Under Hudson River*, 5.

2. Ibid., 6.

3. Ibid., 12.

4. Ibid., 21.

5. Ibid.

6. "Hudson Tube Bill Rejected," *New York Times*, December 13, 1918.

7. Al Smith, *Up to Now: An Autobiography* (New York: Viking, 1920), 4–6, 14–16.

8. Ibid., 158.

9. "15,000 Ordered Out for Harbor Strike," *New York Times*, January 9, 1919; "16,000 Go on Strike and Tie Up Harbor; Break Likely Today," *New York Times*, January 10, 1919; "Federal Effort Fails to Stop Harbor Tieup," *New York Times*, January 11, 1919; "End Harbor Strike on Wilson's Plea," *New York Times*, January 12, 1919.

10. "Edge Urges Vehicle Tube," *New York Times*, January 17, 1919.

11. Jameson W. Doig, *Empire on the Hudson: Entrepreneurial Vision and Politi-

cal Power at the Port of New York Authority (New York: Columbia University Press, 2001), 50.

12. Julius Henry Cohen, *They Builded Better Than They Knew* (New York: Julian Messner, 1946), 117–119; Edge, *Jerseyman's Journal*, 261.

13. Cohen, *They Builded Better Than They Knew*, 117–119; Edge, *Jerseyman's Journal*, 95–96.

14. "Delay Port Treaty for Study by City," *New York Times*, March 2, 1919.

15. "Lesson in Harbor Strike," *New York Times*, March 6, 1919.

16. "Disapprove Port Treaty," *New York Times*, March 13, 1919.

17. Robert A. Caro, *The Power Broker: Robert Moses and the Fall of New York* (New York: Vintage Books, 1975), 96. In a footnote on page 616, Caro claims that Robert Moses, in "one of his first assignments for Al Smith," analyzed two conflicting tunnel-construction proposals: one by the tunnel commission, which would cost twenty-eight million dollars, and the other by Goethals, which would cost twelve million dollars. After interviewing Goethals, the commission's engineers, and independent experts, Moses determined that the Goethals plan would not work and that while the commission's plan would work, the eventual cost would be forty-eight million dollars. Smith, therefore, "threw out the Goethals plan and accepted the commission's, but allocated for it the $48 million that Moses had suggested." If true, Moses's skill in engineering analysis was far greater than that of the best and most experienced tunnel engineers in the country. However, if Moses, as chief of staff of the Reconstruction Commission, conducted his analysis as "one of his first assignments" for Smith, he would have done so in March 1919. At that time, the Goethals's plan *was* the tunnel commission's plan, and the Holland plan, which was later projected to cost twenty-eight million dollars, had not yet been proposed. Moreover, Smith never allocated forty-eight million dollars for the tunnel project, as the New York legislature's appropriations necessary to cover the cost of the tunnel were made over a number of years, some coming after Smith was out of office. Furthermore, New York State was responsible only for half of the construction cost, so Smith would never have had to "allocate" forty-eight million dollars, even if he had the power to do so (which he did not). Caro's source for his information appears to be an interview with Moses conducted decades after the fact and bears signs of Moses's desire to recall events in a manner which flattered his overwhelming ego.

18. "Engineers Study Vehicular Tunnel," *New York Times*, March 16, 1919.

19. "Gov. Smith at City Hall," *New York Times*, March 30, 1919.

20. "Hearing on Tunnel Held by Gov. Smith," *New York Times*, April 6, 1919.

21. Ibid.

22. Ibid.

23. "Smith Signs Tunnel Bill," *New York Times*, April 12, 1919.

24. Clarence C. Smith Correspondence, 1919, Manuscripts and Special Collections, New York State Library, Albany, New York.

25. "Realty Deals Involved in Bitter Fight over Hudson Vehicular Tube," *New*

York Evening Post, May 25, 1921. See example of brochure in Clarence C. Smith Correspondence, 1919, Manuscripts and Special Collections, New York State Library, Albany, New York.

26. Bishop, *Goethals*, 418.

27. *Report of the New York Commission, 1920*, 6.

28. "Vipond Davies May Design the Wagon Tunnel," *Jersey Journal*, April 14, 1919.

29. "J. Vipond Davies; Built Hudson Tube," *New York Times*, October 5, 1939.

30. "Wagon Tunnel Commissioners Hold First Session," *Jersey Journal*, April 25, 1919; "Wagon Tunnel, Goethals and Davies Ideas," *Jersey Journal*, April 26, 1919; "To Pick Engineers," *Jersey Journal*, May 1, 1919; "Tunnel Board at Odds over Plan Engineer," *Jersey Journal*, May 2, 1919.

31. "'Father' of the Wagon Tunnel Is Feasted," *Jersey Journal*, May 31, 1919.

32. "Tunnel Rules That Gen. Goethals Couldn't Accept," *Jersey Journal*, June 5, 1919; "Gen. Goethals Is Out of Wagon Tunnel Board; Holland Put in Charge," *Jersey Journal*, June 5, 1919.

33. "New York Central Changes," *New York Times*, February 1, 1903; "Belt Line Elevated and Subway Project," *New York Times*, March 10, 1909; "William J. Wilgus, Rail Expert, Dead," *New York Times*, October 25, 1949; William J. Wilgus, *Proposed New Railway System for the Transportation and Distribution of Freight by Improved Methods in the City and Port of New York* (New York: Amsterdam, 1908). See also the finding aid and materials of the Wilgus Papers.

34. "William J. Wilgus, Rail Expert, Dead"; Joseph W. Konvitz, "William J. Wilgus and Engineering Projects to Improve the Port of New York, 1900–1930," *Technology and Culture* 30 (1989): 398–425. Konvitz claims that Wilgus "took credit for beginning the process that eventuated in the Port of New York Authority" but was never given credit for this by others (412n. 22). (Wilgus was accorded this credit in his obituary in the *New York Times*.) Unfortunately, Konvitz does not adequately explain why New York never joined with New Jersey in authorizing such a district. In *Empire on the Hudson* (435n. 23), Doig notes Konvitz's article and gives Wilgus credit for influencing the plans of the New York, New Jersey Port and Harbor Development Commission and of the Port Authority. He also notes that the report of the commission mentioned Wilgus's plans for an underground freight-railway system. Not mentioned by Doig, however, is the commission's silence on Wilgus's plans for creation of an interstate metropolitan district. More research into Wilgus's contribution to the concept of a Port Authority for New York and New Jersey is needed.

35. For specific information regarding Wilgus's opposition to a takeover of the tunnel by trucking interests, see "Memo to C. M. Holland, from W. J. Wilgus, March 20, 1920," box 49, Wilgus Papers. For more on Wilgus's proposals for transportation improvements in the Port of New York, see Konvitz, "William J. Wilgus." Konvitz, however, does not mention Wilgus's contributions to the Holland Tunnel. Moreover, he incorrectly states that Wilgus operated on a flawed

assumption about planning—that comprehensive planning for the entire region should precede the execution of any particular project. Wilgus's work on and devotion to the Hudson River vehicular tunnel is evidence to the contrary.

36. "Hudson Under-River Roadway," *New York Times*, June 15, 1919.

37. "Clifford M. Holland, Tunnel Engineer, Dies," *Engineering News Record* 93, no. 18 (October 30, 1924): 723, 726; "Memoir of Clifford Milburn Holland," *Transactions of the American Society of Civil Engineers* 89 (1926): 1625–1629; "Named for Its Engineer, Holland Tunnel Is 40 Years Old Today," *Jersey Journal*, November 13, 1967; *Dictionary of American Biography*, s.v. "Holland, Clifford Milburn."

38. D. A. Gasparini and Judith Wang, "Battery–Joralemon Street Tunnel," *Journal of Performance of Constructed Facilities* 20, no. 1 (February 1, 2006): 104.

39. "In Holland's Tunnel Is His Monument," *New York Times*, November 13, 1927.

40. "Tells Engineer's Dream," *New York Times*, November 12, 1927.

41. C. M. Holland to Colonel W. J. Wilgus, Hannover Inn, Hannover, NH, July 22, 1920, marked "Personal," box 49, Wilgus Papers.

42. "Milton H. Freeman Noted Engineer," *Canton Commercial Advertiser*, March 29, 1925.

43. "Ole Singstad, 87, Master Builder of Underwater Tunnels, Is Dead," *New York Times*, December 9, 1969; Kenneth Bjork, *Saga in Steel and Concrete: Norwegian Engineers in America* (Northfield, MN: Norwegian-American Historical Association, 1947), 188, 189.

44. Ira Wolfert, "Huge Tunnel Built Like a Double-Barreled Shotgun," *Popular Science*, October 1957, 98.

45. "To Begin Work on Wagon Tunnel by January 1 Next," *Jersey Journal*, June 11, 1919; "Wagon Tunnel Commission Doesn't Like Mix-Up over Jobs," *Jersey Journal*, June 18, 1919.

46. Bishop, *Goethals*, 416.

Notes to Chapter 5

1. See document marked "October 2, 1919, Location-Memorandum of Preliminary Studies," box 50, Wilgus Papers.

2. *Report of the New York Commission, 1920*, 76.

3. William J. Wilgus, Chairman, Board of Consulting Engineers, to New York State Bridge and Tunnel Commission and New Jersey Interstate Bridge and Tunnel Commission, January 21, 1920, box 52, Wilgus Papers.

4. "New Vehicular Tube to Cost $28,669,000," *New York Tribune*, February 16, 1920.

5. "Goethals Asks Why His Plan Is Barred," *New York Times*, February 20, 1920.

6. "Byrne's Tunnel Criticism," *New York Times*, February 21, 1920; "Asserts Goethals Tunnel Idea Best," *Sun and New York Herald*, February 21, 1920.

7. "A Highway under the Hudson," *Engineering News-Record* 84, no. 8 (February 19, 1920): 356.

8. The account of this hearing, including all quotations, comes from "Goethals Defends His Tunnel Plan," *New York World*, February 24, 1920.

9. Board of Consulting Engineers, N.Y.-N.J. Interstate Vehicular Tunnel, Meeting Minutes and Reports, Meeting No. 34, March 24, 1920, box 51, Wilgus Papers; C. M. Holland, Chief Engineer, W. J. Wilgus, Chairman, Board of Consulting Engineers, J. A. Bensel, Wm. H. Burr, J. V. Davies to New York State Bridge and Tunnel Commission and New Jersey Interstate Bridge and Tunnel Commission, March 9, 1920, box 52, Wilgus Papers.

10. "Goethals on Jersey Tubes," *New York Times*, March 5, 1920; "Six Lines of Traffic Demanded for Hudson Tunnel," *Engineering News-Record* 84, no. 11 (March 11, 1920): 543.

11. "Arraign Adams, Bar Block Tunnel Plan," *Newark Evening News*, March 10, 1920.

12. "Commissioners Bar Goethals Tunnel," *New York Times*, March 10, 1920.

13. "Oppose Two-Lane Hudson Vehicle Tunnel," *Engineering News-Record* 84, no. 12 (March 18, 1920): 593.

14. In this, the critics were correct. The traffic projections not only underestimated the rate at which use horse-drawn wagons would be superseded by use of motor trucks but also underestimated the extent to which improvements in production efficiency, lowered cost, and aggressive marketing would lead to an explosion in the use of passenger vehicles. Moreover, the engineers did not adequately account for the induced growth in land development in northern New Jersey, which led to a corresponding increase in traffic.

15. "Demand Wider Tunnel," *New York Times*, March 21, 1920.

16. Memorandum explaining the purpose of the Boettger resolution adopted by the New Jersey Interstate Bridge and Tunnel Commission on March 9, 1920, copy of hand-corrected typescript, box 49, Wilgus Papers.

17. "Tube Body Rejects Defense by Adams," *Newark Evening News*, March 17, 1920.

18. "Oppose Two-Lane Hudson Vehicle Tunnel."

19. Draft of Second Tri-Monthly Report of the Board of Consulting Engineers, April 8, 1920, box 49, Wilgus Papers.

20. Walter Gregory Muirheid, *Jersey City of Today: Its History, People, Trades, Commerce, Institutions and Industries* (1910; repr., Englewood, NJ: Bergen, 1996), 66; "John F. Boyle Dead: Was Hague Adviser," *New York Times*, July 3, 1930.

21. For general information on Hague's life and career, see McKean, *The Boss*; Gerald Leinwand, *Mackerels in the Moonlight: Four Corrupt American Mayors* (Jefferson, NC: McFarland, 2004); Richard J. Conners, *A Cycle of Power: The Career of Jersey City Mayor Frank Hague* (Metuchen, NJ: Scarecrow, 1971); and Thomas Fleming, "I Am the Law," *American Heritage* 20 (June 1969): 32–48.

22. Fleming, "I Am the Law."

23. Thomas Fleming, *Mysteries of My Father* (Hoboken, NJ: Wiley, 2005), 156.

24. "Statement by Commissioner T. Albeus Adams at the meeting of the New Jersey Interstate Bridge and Tunnel Commission, held May 4, 1920, after the consideration of Tri-Monthly Report No. 2 Submitted by the Board of Consulting Engineers," box 49, Wilgus Papers.

25. Ibid.

26. Ibid.

27. "Vehicle Tube Bill Passed," *New York Times*, May 12, 1920; "Signs Bill to Begin Jersey Tunnel," *New York Times*, May 26, 1920.

28. *Engineering News-Record* 84, no. 24 (June 3, 1920).

Notes to Chapter 6

1. E. M. Barradale, Assistant to the Chief Engineer, to Mr. C. M. Holland, Chief Engineer, New York and New Jersey Bridge and Tunnel Commissions, Woolworth Building, New York, December 6, 1923, box 51, Wilgus Papers.

2. Board of Consulting Engineers, N.Y.-N.J. Interstate Vehicular Tunnel, Meeting Minutes and Reports, Meeting No. 48. September 22, 1920, box 51, Wilgus Papers.

3. William J. Wilgus to Mr. T. A. Adams, Chairman, New Jersey Interstate Bridge and Tunnel Commission, 616 Hall of Records, New York City, May 10, 1921, box 51, Wilgus Papers.

4. "Vehicle Tube Held Up While Officials Fight," *New York Evening Post*, May 26, 1921.

5. Fleming, *Mysteries of My Father*, 156.

6. C. M. Holland, Chief Engineer, to Col. W. J. Wilgus, Consulting Engineer, 165 Broadway, New York City, May 23, 1921, box 49, Wilgus Papers.

7. Memorandum of Negotiations with Representatives of Erie Railroad (Mr. Falconer and Mr. King) and Representatives of Committee of the Vehicular Tunnel Commissions (W. J. Wilgus, C. M. Holland, and J. B. Snow), n.d., box 50, Wilgus Papers.

8. C. M. Holland to Colonel W. J. Wilgus, Hanover Inn, Hanover, N.H., July 22, 1920, box 49, Wilgus Papers.

9. J. A. L. Waddell, "Bridge versus Tunnel for the Proposed Hudson River Crossing at New York City," *Transactions of the American Society of Civil Engineers* 84 (1921): 571.

10. C. M. Holland, J. A. Bensel, William J. Wilgus, J. Vipond Davies, Geo. S. Watson, William H. Burr, Frank M. Williams, J. B. Snow, M. H. Freeman, O. Singstad, R. Smillie, Orrin L. Brodie, and O. F. Bellows to Publication Committee, American Society of Civil Engineers, 29 West 39th Street, New York City, August 23, 1920, box 49, Wilgus Papers.

11. "Discussion," of Waddell, "Bridge versus Tunnel," *Transactions of the American Society of Civil Engineers* 84 (1921): 576.

12. See New Jersey Interstate Bridge and Tunnel Commission, *Report of the New Jersey Interstate Bridge and Tunnel Commission to the Senate and General Assembly of New Jersey, 1921* (Trenton, NJ: MacCrellish and Quigley, 1921), Appendix No. 3 and Appendix No. 4; "Research Reveals How to Ventilate the Hudson Tunnel," *Compressed Air Magazine*, April 1922, 101–107.

13. *Report of the New Jersey Commission, 1921*, 102.

14. Ibid.

15. "First Bids Opened for Vehicular Tube," *New York Times*, September 22, 1920; "Many Lower Greenwich Village Landmarks Doomed by the Jersey Vehicular Tunnel," *New York Times*, October 31, 1920.

16. Descriptions of the ceremony are taken from a number of sources, including "Columbus Day to See the Start of Twin-Tube Thoroughfare Which Will Join New York and New Jersey beneath the River," *New York World*, October 10, 1920; "Governors to Start Work on Jersey Tubes," *New York Times*, October 11, 1920; "State Heads Dig Jersey Tube Hole," *New York World*, October 13, 1920; "Ground Broken for Vehicular Tunnel," *New York Herald*, October 13, 1920; and "Ground Is Broken for Vehicular Tube," *New York Times*, October 13, 1920. Following construction of the tunnel, this park was taken over by the city sanitation department and not returned to park use until 2005.

17. "College to Study Tunnel Air Supply," *New York Times*, January 26, 1921; "Study Tube Ventilation," *New York Times*, April 3, 1921.

18. "Vehicle Tube Held Up While Officials Fight," *New York Evening Post*, May 26, 1921; see also "Gen. Dyer Charges Tunnel Deadlock up to Jersey Board," *Jersey Journal*, May 28, 1921.

Notes to Chapter 7

1. George L. Watson, M. Am. Soc. C.E., Consulting Engineer, 141 Broadway, New York City, N.Y., to the Board of Consulting Engineers, New York and New Jersey Bridge and Tunnel Commission, Hall of Records, New York City, N.Y., March 5, 1921, Board of Consulting Engineers Minute Book, box 52, Wilgus Papers.

2. George L. Watson to the New York State Bridge and Tunnel Commission and the New Jersey Interstate Bridge and Tunnel Commission, Hall of Records, New York City, N.Y., March 14, 1921, Board of Consulting Engineers Minute Book, box 52, Wilgus Papers.

3. "Vehicular Tunnel Hits Serious Snag," *New York Times*, April 13, 1921.

4. "Hague Demands Pact to Widen Streets Be Kept," *Jersey Journal*, April 26, 1922.

5. "New Tunnel Work Delayed," *New York Times*, March 24, 1921.

6. "Tube Row Referred to State Attorney," *New York Times*, May 4, 1921.

7. "English Auto Club an Example Here," *New York Times*, December 31, 1913; "Would Post Notice about Auto Fines," *New York Times*, January 26, 1914; "How 'Lincoln Way' Project Now Stands," *New York Times*, April 5, 1914; "A Famous Highway," *New York Times*, July 26, 1914.

8. *Jersey Journal*, May 4, 1921.

9. "Tunnel Board Pleased That Watson Quit," *Hudson Dispatch*, May 19, 1921.

10. "Seek to Prevent Tunnel Board Agreement," *Jersey Journal*, May 25, 1921.

11. "Realty Deals Involved in Bitter Fight over Hudson Vehicular Tube," *New York Evening Post*, May 25, 1921.

12. Ibid.

13. "Vehicle Tube Held Up While Officials Fight," *New York Evening Post*, May 26, 1921.

14. "Edwards Defends Tunnel Chairman," *New York Times*, May 27, 1921; see also "Highway Board Here June 22 to Plan Tunnel Approaches," *Jersey Journal*, May 27, 1921.

15. "Tunnel Fight Ends; Ask Bids Tuesday," *New York Times*, June 8, 1921; "Agree to Street Widenings at Jersey City End," *Jersey Journal*, June 8, 1921.

16. "Tunnel Deadlock Ended," *Jersey Journal*, June 8, 1921.

17. "Engineers Censure Tunnel Consultants," *Newark Evening News*, July 15, 1921.

18. "Plan Tunnel Celebrations," *New York Times*, July 17, 1921.

19. "New Bids Received for Hudson Vehicle Tunnel," *Engineering News-Record* 84, no. 3 (July 21, 1921): 124.

20. Memorandum of Conference in Chief Engineer's Office on Contract No. 2, July 21, 1921, box 51, Wilgus Papers; Report on Bids, Contract No. 2, July 26, 1921, to New York State Bridge and Tunnel Commission and New Jersey Interstate Bridge and Tunnel Commission, from Chief Engineer, box 51, Wilgus Papers.

21. "Bidder Told $14,000 of Contract Went to Connolly Fund," *New York Times*, October 7, 1921.

22. "Jersey Tube Work Is Again Tied Up," *New York Evening Post*, September 22, 1921.

23. "Engineers to Test Hudson Tube Model," *New York Times*, June 15, 1921; Robert G. Skerrett, "Research Reveals How to Ventilate the Hudson Tunnel," *Compressed Air Magazine*, April, 1922, 106.

24. "Tests Show Safety of Vehicular Tunnel," *New York Times*, October 30, 1921; New Jersey Interstate Bridge and Tunnel Commission, *Report of the New Jersey Interstate Bridge and Tunnel Commission to the Senate and General Assembly of the State of New Jersey, 1922* (Union Hill, NJ: Hudson, 1922), 46–49 (hereafter *Report of the New Jersey Commission, 1922*).

25. "Tests Show Safety of Vehicular Tunnel."

26. "5 Jersey Cities Ask Inquiry on Tunnel," *New York Times*, December 9, 1921.

27. "Snub for Jersey's Tunnel Commission," *New York Times*, December 15, 1921; "Jersey Wants Memorial," *New York Times*, December 21, 1921.

28. "5 Jersey Cities Ask Inquiry on Tunnel."

29. "Open Tunnel Bids Today," *New York Times*, February 7, 1922.

30. "$19,250,000 Lowest Jersey Tunnel Bid," *New York Times*, February 16, 1922.

31. "Tunnel Funds Wait Award of Contract," *New York Times*, February 10, 1922; "$19,250,000 Lowest Jersey Tunnel Bid."

Notes to Chapter 8

1. "See Adams' Hand in Tunnel Bill," *New York Evening Post*, January 30, 1922.

2. "Shift Follows Declaration of War by Hague," *Jersey Journal*, February 2, 1922.

3. Ibid.

4. "State Tunnel Body Blamed for Delay," *New York Times*, February 24, 1922.

5. The following account of the General Assembly session, including all quotations, comes from "Tunnel Board in Jersey Doomed by Vote in Assembly," *New York Evening Post*, March 7, 1922.

6. "Harmony Prevails as New N.J. Bridge Board Organizes," *New York Tribune*, March 22, 1922.

7. "New Tunnel Board, with Friction Over, Tackles Big Job," *Jersey Journal*, March 21, 1922; "New Commission Asks Mayor to Give Up Streets," *Jersey Journal*, March 22, 1922; "Tunnel Award Today," *New York Times*, March 28, 1922; "Contract Awarded for Jersey Tunnel," *New York Times*, March 29, 1922; "Awards Contract for Jersey Tube," *New York World*, March 29, 1922; "Hudson River Tube Dirt Flies Today," *New York World*, March 31, 1922.

8. "Hague Firm as Tunnel Board Adopts Ultimatum," *Jersey Journal*, April 5, 1922.

9. "Condemnation May End Tunnel Delay," *New York Times*, April 23, 1922.

10. "Hague Demands Pact to Widen Streets Be Kept," *Jersey Journal*, April 26, 1922; "Hudson Tunnel Delayed," *New York Times*, April 27, 1922.

11. "Tunnel Board Hesitates as to Eleventh Street Grab," *Jersey Journal*, May 10, 1922.

12. "Tunnel Is Started Secretly in Jersey," *New York Times*, June 1, 1922; "Ground Broken at Jersey End of Hudson Vehicular Tunnel," *Engineering News-Record* 88, no. 157 (June 8, 1922): 157.

13. "Silent on Tunnel Work," *New York Times*, June 2, 1922.

14. "Condemning Land for New Tunnels," *New York Times*, July 23, 1922.

15. McKean, *The Boss*, 71, 78.

16. "Court Rules Jersey City Can't Supervise Tunnel Job," *Jersey Journal*, July 13, 1922; "Jersey City Loses Fight on Tunnel," *New York Times*, July 14, 1922.

17. "Tunnel Work Going Well, Say Officials," *New York Times*, August 9, 1922.

18. *Report of the New Jersey Commission, 1922*, 27.

19. "The Successful Sinking of Two Great Caissons," *Compressed Air Magazine*, February 1922, 41–43; *Report of the New Jersey Commission, 1922*, 24.

20. "Tunnel Work Going Well, Say Officials."

21. "Driving Our First Motor Highway under Hudson Sandhog's Biggest Joy," *New York Times*, December 10, 1922.

22. "Trip of Big Caisson Stiff Job for Tugs," *New York Times*, January 31, 1923.

Notes to Chapter 9

1. "A Substitute for Sand-Hog," *New York Times*, April 22, 1923.

2. *Missouri Republican*, March 25, 1870; Robert W. Jackson, *Rails across the Mississippi: A History of the St. Louis Bridge* (Urbana: University of Illinois Press, 2001) 82, 83.

3. "Coroner Will Probe River Tunnel Deaths," *New York Times*, June 5, 1906; " 'The Bends' Hit Scores in Pennsylvania Tubes," *New York Times*, June 16, 1906.

4. "2 Drowned in Tunnel," *New York Times*, June 21, 1906.

5. B. H. M. Hewett and S. Johannesson, *Shield and Compressed Air Tunneling* (New York: McGraw-Hill, 1922), 431. These numbers represent the "official" statistics and do not include injuries or deaths from other causes.

6. Paul E. Delaney, *Sandhogs: A History of the Tunnel Workers of New York* (New York: Longfield, 1983), 33.

7. "13 Gave Lives for Tunnel," *Jersey Journal*, November 21, 1927.

8. "Live-Saving 'Air-Locks,'" *New York Times*, December 21, 1921; "Tunnel 'Sand-Hogs' Go Out on Strike," *New York Times*, April 13, 1923.

9. " 'Sandhog' Strike Settled," *New York Times*, April 17, 1923.

10. "Killed by Fall into Caisson of New Hudson River Tube," *New York Times*, August 21, 1923; "13 Gave Lives for Tunnel."

11. "Radio Waves Heard in the Jersey Tube," *New York Times*, January 5, 1924.

12. New Jersey Interstate Bridge and Tunnel Commission, *Report of the New Jersey Interstate Bridge and Tunnel Commission to the Senate and General Assembly of the State of New Jersey, 1924* (Trenton, NJ: MacCrellish and Quigley, 1924), 8.

13. "Smith Asks Jersey to Merge on Tube," *New York Times*, March 6, 1923; "Silzer Urges Plan of Port Authority," *New York Times*, January 23, 1924; "The Vehicular Tunnel," *New York Times*, January 23, 1924; Frederick L. Bird, *A Study of the Port of New York Authority* (New York: Dun and Bradstreet, 1945), 13; Doig, *Empire on the Hudson*, 144.

14. "Discuss Bond Issue to Complete Tunnel," *New York Times*, January 29, 1924; "Port Authority Control of Tunnel Is Opposed," *New York Times*, February 10, 1924.

15. "Vehicular Tunnel Work Progressing," *New York Times*, March 2, 1924.

16. " 'Sandhogs' Strike in Hudson Tunnel," *New York Times*, April 10, 1924.

17. "Some 'Sandhogs' Return," *New York Times*, April 11, 1924.

18. "Auto Fumes Poison Scores in Tunnel," *New York Times*, May 11, 1924; "Pittsburgh Fears Trolley Strike Riots," *New York Times*, May 12, 1924.

19. "Auto Fumes Poison Scores in Tunnel."

20. A list of these slides may be found in the C. M. Holland Collection, Special Collections, Kelvin Smith Library, Case Western Reserve University.

21. "Tells Engineer's Dream," *New York Times*, November 12, 1927.

22. "Holland Tunnel," *Time*, November 21, 1927.

23. "Tells Engineer's Dream."

24. "Tubes under River Will Meet Oct. 29," *New York Times*, October 12, 1924; "Blast Joins Halves of Vehicular Tube: Meet within an Inch," *New York Times*, October 30, 1924.

25. "In Holland's Tunnel Is His Monument," *New York Times*, November 13, 1927.

26. "'Sandhog' Says He Was First to Go through Holland Tubes," *New York Times*, November 13, 1927.

27. "Blast Joins Halves of Vehicular Tube: Meet within an Inch."

Notes to Chapter 10

1. "13 Gave Lives for Tunnel," *Jersey Journal*, November 12, 1927.

2. New Jersey Interstate Bridge and Tunnel Commission, *Report of the New Jersey Interstate Bridge and Tunnel Commission to the Senate and General Assembly of the State of New Jersey, 1926* (Trenton, NJ: MacCrellish and Quigley, 1926), 37 (hereafter *Report of the New Jersey Commission, 1926*).

3. "$13,500,000 Highway on West Side Voted," *New York Times*, June 15, 1926; "Strongly Opposes Elevated Highway," *New York Times*, May 12, 1927; "Walker Beats Move by Miller to Push Express Highway," *New York Times*, April 5, 1927.

4. "5,000 in the Village Defy Order to Move," *New York Times*, August 31, 1926; "$4,500,000 Awarded in Sixth Avenue Extension," *New York Times*, March 25, 1927; "Says Trucking Firm Blocks Tube Outlet," *New York Times*, October 31, 1927.

5. "5,000 in the Village Defy Order to Move."

6. Ibid.

7. "Many Lower Greenwich Village Landmarks Doomed by the Jersey Vehicular Tunnel," *New York Times*, October 31, 1920.

8. Ibid.

9. "Fight Plaza Change in River Tube Plan," *New York Times*, February 24, 1926; "Says Trucking Firm Blocks Tube Outlet."

10. "J. J. Riordan Ends Life with Pistol in His Home; His Bank Declared Sound," *New York Times*, November 10, 1929.

11. "Hague Plan Wins Favor," *New York Times*, September 21, 1926.

12. "Great Express Highways for New York Zone," *New York Times*, November 21, 1926; Hart, *Last Three Miles*, 73.

13. "Smith and Moore Meet in Tunnel," *New York Times*, August 22, 1926; "Governors Join Hands Where Tubes Join States," *Jersey Journal*, August 23, 1926.

14. "Contractor Drowned in Vehicular Tube after Drop of 80 Feet in Dark into Pool," *New York Times*, December 21, 1926. The death of the unnamed sandhog who died in 1925 from pneumonia while being treated in the hospital for caisson disease (see page 37 of *Report of the New Jersey Commission, 1926*) combined with the deaths of the thirteen workers officially listed as tunnel fatalities and the deaths of Holland and Freeman totals sixteen deaths which may justly be attributed to the tunnel project. In 1927, Martin Casey, one of the sandhogs who worked on the project, stated, "There were only sixteen men killed. That is a pretty low score for a job like that. Of course, the engineers deserve all the glory they get, but down in the hole, there are no white collar jobs." (See "Sandhogs Too Busy Digging to Attend Tunnel Opening," *Jersey Journal*, November 14, 1927.) It is unclear if Casey included the two white-collar engineers in his total or if he was speaking only of laborers.

15. "Holland Tunnels to Cost $48,400,000," *New York Times*, December 27, 1926.

16. "Gas Bomb Fumes Test Holland Tube," *New York Times*, March 16, 1927.

17. "Ventilation in Tube Called a Menace," *New York Times*, April 14, 1927.

18. "Urges Fire Test in Tube," *New York Times*, April 22, 1927.

19. "300 Traffic Police for Holland Tube," *New York Times*, May 16, 1927.

20. "Automobile Burned in New Tube as Test," *New York Times*, November 4, 1927.

21. "Low Tunnel Rate for Buses Asked," *New York Times*, October 26, 1927.

22. "Fears Truck Jam in Holland Tunnel," *New York Times*, September 17, 1927; "How It Feels to Ride in the Holland Tunnel," *New York Times*, October 2, 1927.

22. "Toll Scale Is Fixed for Holland Tunnel," *New York Times*, November 2, 1927.

24. "Truckmen Protest Holland Tube Rates," *New York Times*, November 3, 1927.

25. "Ferries Big Factor in New Tube Tolls," *New York Times*, November 5, 1927.

Notes to Chapter 11

1. "Ole Singstad, 87, Master Builder of Underwater Tunnels, Is Dead," *New York Times*, December 9, 1969.

2. The events that occurred on opening day and the days immediately preceding and following the opening were covered by numerous articles in a special commemorative edition of the *Jersey Journal* on November 12, 1927; by editions of the *Jersey Journal* on November 13 and 14, 1927; and by editions of the *New York Times* on November 13 and 14, 1927.

3. "Tunnel Fete Sidelights," *Jersey Journal*, November 14, 1927.

4. "Great Crowd Treks into Holland Tubes after Gala Opening," *New York Times*, November 13, 1927; "More Tunnels Seen as Throngs Hail Twin Bores," *Jersey Journal*, November 14, 1927. Goethals died on January 21, 1928.

5. "Holland Tunnel Will Be 50 Years Old on November 12, 1977," news release by Public Affairs Department, Port Authority of New York and New Jersey, October 31, 1977, Holland Tunnel vertical file, Jersey City Free Public Library, Jersey City, New Jersey.

6. "Traffic in Tunnel 40 Percent Trucks," *New York Times*, November 15, 1927.

7. "Rail Ferries Hurt by Holland Tunnel," *New York Times*, December 20, 1927.

8. "Tunnel Competition Cuts Ferry Rates," *New York Times*, May 8, 1928.

9. "Holland Tunnel Traffic Gains Steadily; 24,384 Vehicles a Day Used Tube in May," *New York Times*, June 11, 1928.

10. New York State Bridge and Tunnel Commission, "Statement of Total Traffic, Operation of the Holland Tunnel, 1930," reproduced from typewritten copy, Science, Industry and Business Library, New York Public Library.

11. "Our New Tube Is a 'Sight,'" *New York Times*, February 5, 1928.

12. New York State Bridge and Tunnel Commission, "Statement of Total Traffic, Operation of the Holland Tunnel from November 13, 1927, to January 31st, 1929," document located in the Science, Industry and Business Library, New York Public Library.

13. "Tunnel Buses Cut Tube Train Travel," *New York Times*, March 15, 1929; "Ferry Traffic Cut 3,246,881 in Year," *New York Times*, April 24, 1929.

14. "Abell's Commission Begins Inquiry Today," *New York Times*, September 23, 1929; "Holland Tunnel High Salaries Bring Investigation," *Jersey Journal*, December 10, 1929; "Critical of Pay in Holland Tunnel," *New York Times*, December 11, 1929; "Holland Tunnel Legislation Investigator Kimberly Claims $300,000 Annual Waste in Operation," *Jersey Journal*, January 7, 1930.

15. "Holland Tube Waste Denied by Officials," *New York Times*, January 14, 1930.

16. Ibid.

17. The Holland Society of New York was organized in 1885 to collect information regarding the early Dutch settlement of New Netherland and has nothing to do with Clifford Holland or the Holland Tunnel. "Roosevelt Urges Uniting Port Bodies," *New York Times*, January 17, 1930; "Two States Agree on New Hudson Tube," *New York Times*, February 15, 1930. The circumstances of the consolidation are well summarized in Doig, *Empire on the Hudson*, 169–171.

18. In addition to Dyer and Shamberg, John J. Pulleyn was a carryover from the tunnel commissions representing New York. The carryover commissioners representing New Jersey were Joseph G. Wright of Patterson, Ira R. Crouse of Perth Amboy, and George de B. Keim of Edgewater. "J. F. Galvin Heads Port Authority," *New York Times*, April 25, 1930. See also "Tunnel Program Speeded at

Albany," *New York Times*, April 9, 1930; "Port Authority Bills Signed by Roosevelt," *New York Times*, April 14, 1930; "Port Merger Bills Passed in Jersey," *New York Times*, April 16, 1930.

19. Bird, *Study of the Port of New York Authority*, 18.

20. "Holland Tunnel, Used by 31,000,000 Vehicles, Is 3 Years Old Today; Income Gain in Prospect," *New York Times*, November 13, 1930; "35,000,000 Vehicles Have Used Holland Tunnel since Opening," *New York Times*, March 20, 1931; "Pays Back $50,000,000 on Holland Tunnels," *New York Times*, March 22, 1931.

21. Doig, *Empire on the Hudson*, 184–185.

22. "Express Road Unit Opened by Miller," *New York Times*, November 14, 1930.

23. "Jersey Auto Viaduct Dedicated Today," *New York Times*, November 23, 1932; "New Viaduct a Time-Saver," *New York Times*, November 20, 1932.

24. "39th St. Tube Gets Name of Lincoln," *New York Times*, April 17, 1937.

25. In *The Power Broker: Robert Moses and the Fall of New York* (608), Robert Caro details the differences of opinion between Singstad and Moses regarding how tunnels should be designed and built, and he notes how Singstad "despised Moses' aides, who he felt violated engineering ethics every time they subordinated their own objective views of an engineering problem to tell the Triborough chairman what he wanted to hear."

26. "One Killed, 2 Hurt in Holland Tube Collision: First Fatal Crash Halts Traffic 20 Minutes," *New York Times*, March 19, 1932.

Notes to Chapter 12

1. The details of this story are taken mainly from contemporary accounts in the *New York Times* and the *Jersey Journal* and from a 1949 report produced by the National Board of Fire Underwriters. Accounts of the incident and its aftermath vary considerably. For example, Tyndall is reported in the *Times* as being thirty-nine years old, in one *Journal* article as being thirty-six years old, and in another as being forty years old. The version of the story presented herein is the author's best guess of what happened.

2. National Board of Fire Underwriters, *The Holland Tunnel Chemical Fire: New Jersey–New York, May 13, 1949* (New York: National Board of Fire Underwriters, 1949), 6.

3. *Boyce Motor Lines, Inc. v. United States*, 342 U.S. 337 (1952).

4. "After the Explosion," *New York Times*, May 18, 1949.

5. National Board of Fire Underwriters, *Holland Tunnel Chemical Fire*.

6. "Holland Tunnel Traffic Is Tied Up 4 Hours When Big Truck Overturns and Catches Fire," *New York Times*, January 3, 1946.

7. Amy Allen, letter to the editor, *New York Times*, June 2, 1949.

8. "Tunnel Blast," *Jersey Journal*, May 14, 1949; Alexander Feinberg, "Tunnel

Repairs Speeded: Reopening Today Sought," *New York Times*, May 15, 1949; "Day Traffic Underway in Tunnel," *Jersey Journal*, May 16, 1949; National Board of Fire Underwriters, *Holland Tunnel Chemical Fire*, 4. Tyndall reported to the *Jersey Journal* that he was about one thousand feet into the tunnel when the explosion occurred, but the National Board of Fire Underwriters investigation found that his truck was about twenty-nine hundred feet into the tunnel when it came to rest. Given one eyewitness account that stated that the truck swerved in and out of the slow and fast lanes before coming to a stop, the estimate contained herein is based on the author's conjecture of where the explosion probably occurred.

9. Irving Spiegel, "Chaotic Scenes in Tunnel Described by the Injured," *New York Times*, May 14 1949.

10. Ibid.

11. Alexander Feinberg, "Phone Circuits Cut," *New York Times*, May 14, 1949.

12. "Eye Witness Tells of Blast and Escapes from Tunnel," *Jersey Journal*, May 13, 1949.

13. Spiegel, "Chaotic Scenes in Tunnel."

14. Ibid.

15. It is possible that Lushkin was the one who turned in the first fire alarm, but accounts are unclear and contradictory.

16. National Board of Fire Underwriters, *Holland Tunnel Chemical Fire*, 4.

17. Ibid. 7.

18. "Oxygen Key Factor in Tunnel Fire," *New York Times*, May 14, 1949.

19. Spiegel, "Chaotic Scenes in Tunnel."

20. "Oxygen Key Factor in Tunnel Fire."

21. "Smoke Poisoning Takes Heavy Toll," *New York Times*, May 14, 1949.

22. "Blast Cuts Calls to West and South," *New York Times*, May 14, 1949.

23. Spiegel, "Chaotic Scenes in Tunnel."

24. National Board of Fire Underwriters, *Holland Tunnel Chemical Fire*, 12.

25. Spiegel, "Chaotic Scenes in Tunnel."

26. Ibid.; National Board of Fire Underwriters, *Holland Tunnel Chemical Fire*, 10.

27. Alexander Feinberg, "Wrecked Tunnel Reopened 56 Hours after Blast Here," *New York Times*, May 16, 1949.

28. "Staff Commended by Port Authority," *New York Times*, May 18, 1949.

29. "It's a Brand-New Construction Job Every Night in the Holland Tunnel," *Engineering News-Record* 143, no. 3 (July 21, 1949): 32–34; "Holland Tube Repairs Nearly Done: Full Service Promised for Holiday," *New York Times*, August 30, 1949.

30. Feinberg, "Tunnel Repairs Speeded."

31. "Authority Fears New Tunnel Blasts," *New York Times*, October 21, 1949.

32. National Board of Fire Underwriters, *Holland Tunnel Chemical Fire*, 1.

33. "Jersey Acts to End Truck Cargo Perils," *New York Times*, June 1, 1949; "Action on the Tunnel," *New York Times*, June 2, 1949.

34. "Again, the Tunnel," *New York Times,* June 18, 1949.

35. National Board of Fire Underwriters, *Holland Tunnel Chemical Fire,* 17, 18.

36. "Company Indicted for Tunnel Blast," *New York Times,* October 5, 1949.

37. "Chemical Firm Indicted in Holland Tunnel Blast," *New York Times,* January 10, 1950.

38. "Authority Gets $300,000," *New York Times,* June 9, 1950.

39. "Review Set on Blast in Holland Tunnel," *New York Times,* October 16, 1951.

40. "Truck Company Upheld," *New York Times,* June 17, 1950; *Boyce Motor Lines, Inc. v. United States,* 342 U.S. 337 (1952).

41. "To Honor Heroic Firemen," *New York Times,* June 18, 1950.

42. *Boyce Motor Lines, Inc. v. United States,* 342 U.S. 337 (1952).

43. Ibid.

44. "Truck Line Fined $8,000," *New York Times,* May 20, 1952.

45. "Motorists Flee Tunnel Fire: Fog Causes Highway Jams," *Jersey Journal,* November 14, 1951; "It Was a Bad Day for Commuters, What with Fog, Tube Fire, 'Bomb,'" *New York Times,* November 16, 1951.

46. "Blaze in Truck Ties Up Holland Tunnel Traffic," *New York Times,* June 15, 1964.

Notes to Chapter 13

1. Doig, *Empire on the Hudson,* 145.

2. Ibid., 146.

3. Ibid., 145.

4. "Port Shakeup Fails as Two Boards Merge," *New York Times,* May 2, 1930.

5. Jameson W. Doig, "Joining New York City to the Greater Metropolis: The Port Authority as Visionary, Target of Opportunity, and Opportunist," in *The Landscape of Modernity: Essays on New York City, 1900–1940,* ed. David Ward and Oliver Zunz (New York: Russell Sage Foundation, 1992), 94, 95.

6. Doig, *Empire on the Hudson,* 8.

7. "Tunnel Increased New Jersey Values," *New York Times,* April 30, 1930.

8. Robert M. Fogelston, *Downtown: Its Rise and Fall, 1890–1950* (New Haven: Yale University Press, 2001), 194–199.

9. What Michael Fein has found to be true in regard to the economic impacts of the New York State Thruway and the Tappan Zee Bridge across the Hudson River may also be said of the Holland Tunnel. "Not all economic change," Fein points out, "was a straightforward 'market response' to the improved highway system." Michael R. Fein, *Paving the Way: New York Road Building and the American State, 1880–1956* (Lawrence: University Press of Kansas, 2008), 216.

10. "Varick Street Industrial Center Is Expanding Rapidly," *New York Times,* November 10, 1929.

11. "Funds Approved for Tunnel Links," *New York Times,* October 12, 1951.

12. "Shell of the Third Lincoln Tunnel to Be Completed within 60 Days," *New York Times*, March 14, 1956; "New Lincoln Tunnel Expressway to Alleviate Midtown Congestion," *New York Times*, February 19, 1957; "Tube Connection Will Open Today," *New York Times*, February 19, 1957; "3rd Lincoln Tube Is Opened: Big Test Due over Holiday," *New York Times*, May 26, 1957.

13. Joseph C. Ingraham, "Third Tube Proposed by Moses," *New York Times*, July 2, 1956.

14. Joseph C. Ingraham, "Widening a Tunnel," *New York Times*, August 12, 1956.

15. "Pioneer Designer of Tunnels Here Bemoans All Those Cars," *New York Times*, July 19, 1967; "Ole Singstad, 87, Master Builder of Underwater Tunnels, Is Dead," *New York Times*, December 9, 1969.

16. *Report of the New York Tunnel Commission, 1920*, 42.

17. "Holland Tunnel Starts Its 31st Year of Traffic," *New York Times*, November 13, 1957.

18. Bruce E. Seely, *Building the American Highway System: Engineers as Policy Makers* (Philadelphia: Temple University Press, 1987), 6. For a more recent expression of Seely's argument, see Mark H. Rose, Bruce E. Seely, and Paul F. Barrett, *The Best Transportation System in the World: Railroads, Trucks, Airlines, and American Public Policy in the Twentieth Century* (Columbus: Ohio State University Press, 2006). Seely and his associates have been concerned primarily not with the creation of urban infrastructure by local authorities but with the formation of federal and state policy in regard to long-distance highways constructed with federal-aid funds. Therefore, his studies have limited application to projects such as the Holland Tunnel. One note of caution in the universal application of Seely's theories may be found in Fein, *Paving the Way* (79). Although Fein accepts Seely's basic argument and finds that the road-building regime in New York State in the 1920s was not an automatic response to shifts in technology, he also admits that "the impact of rising automobile use on highway politics during this period is hard to underestimate, as New York motor vehicle registrations doubled from one million to two million in the five years between 1922 and 1927."

19. "Ole Singstad, 87, Master Builder."

20. Konvitz, "William J. Wilgus," 418, 419; Fogelston, *Downtown*, 260.

21. "William J. Wilgus, Rail Expert, Dead," *New York Times*, October 25, 1949.

22. "T. A. Adams Dead: Retired Executive," *New York Times*, September 15, 1940.

23. Doig, *Empire on the Hudson*, 468n. 75.

24. Hart, *Last Three Miles*, 176.

25. "Holland Tubes Run by 'Electronic Brain,'" *New York Times*, March 22, 1928.

26. "Release Two Japs Who Took Tunnel Photos," *Jersey Journal*, September 3, 1935; "Shipping News and Notes," *New York Times*, July 29, 1950.

27. "# Held Overseas in Plot to Bomb Hudson Tunnels," *New York Times*, July 8, 2006.

28. "Bids $12,000,000 on Hudson Truck Tube," *New York Times*, December 15, 1917.

29. "Holland Tunnel, on Eve of 50th Birthday, Is Still Packing Them In," *New York Times*, November 11, 1977.

30. "Senator Edge Makes Speech," *Jersey Journal*, November 14, 1927.

Bibliography

Manuscript Collections

Clarence C. Smith Correspondence, 1919, Manuscripts and Special Collections, New York State Library, Albany, New York

C. M. Holland Collection, Special Collections, Kelvin Smith Library, Case Western Reserve University

O'Rourke Engineering & Construction Company, Heller Transportation Collection, Long Island University, Brooklyn Campus

William J. Wilgus Papers, Manuscripts and Archives Division, New York Public Library, Astor, Lenox and Tilden Foundations

Newspapers

Canton Commercial Advertiser (NY)
Hudson Dispatch (NJ)
Jersey Journal (NJ)
Missouri Republican
Newark Evening News
New York Evening Post
New York Herald
New York Times
New York Tribune
New York World
Sun and New York Herald

Public Documents

U.S. Congress, Senate Committee on Interstate Commerce. *Tunnel Under Hudson River: Hearing on S. 4765.* 65th Cong., 3d sess., December 12, 1918.

U.S. Department of Transportation, Federal Highway Administration. "Highway Statistics Summary to 1995." http://www.fhwa.dot.gov/ohim/summary95/mv 200.pdf.

Books, Articles, and Published Reports

Bard, Erwin W. *The Port of New York Authority.* New York: Columbia University Press, 1942.

Bianculli, Anthony J. *Trains and Technology: The American Railroad in the Nineteenth Century.* Vol. 4, *Bridges and Tunnels, Signals.* Newark: University of Delaware Press, 2003.

Bird, Frederick L. *A Study of the Port of New York Authority.* New York: Dun and Bradstreet, 1945.

Bishop, Joseph Bucklin. *Goethals: Genius of the Panama Canal; A Biography.* New York: Harper and Brothers, 1930.

Bjork, Kenneth. *Saga in Steel and Concrete: Norwegian Engineers in America.* Northfield, MN: Norwegian-American Historical Association, 1947.

Caro, Robert A. *The Power Broker: Robert Moses and the Fall of New York.* New York: Vintage Books, 1975.

Cohen, Julius Henry. *They Builded Better Than They Knew.* New York: Julian Messner, 1946.

Condit, Carl W. *The Port of New York: A History of the Rail and Terminal System from the Beginnings to Pennsylvania Station.* Chicago: University of Chicago Press, 1980.

Conners, Richard J. *A Cycle of Power: The Career of Jersey City Mayor Frank Hague.* Metuchen, NJ: Scarecrow, 1971.

Delaney, Paul E. *Sandhogs: A History of the Tunnel Workers of New York.* New York: Longfield, 1983.

Doig, Jameson W. *Empire on the Hudson: Entrepreneurial Vision and Political Power at the Port of New York Authority.* New York: Columbia University Press, 2001.

———. "Joining New York City to the Greater Metropolis: The Port Authority as Visionary, Target of Opportunity, and Opportunist." In *The Landscape of Modernity: Essays on New York City, 1900–1940,* edited by David Ward and Oliver Zunz, 76–105. New York: Russell Sage Foundation, 1992.

Edge, Walter Evans. *A Jerseyman's Journal: Fifty Years of American Business and Politics.* Princeton: Princeton University Press, 1948.

Fein, Michael R. *Paving the Way: New York Road Building and the American State, 1880–1956.* Lawrence: University Press of Kansas, 2008.

Finch, Christopher. *Highways to Heaven: The Auto Biography of America.* New York: HarperCollins, 1992.

Fleming, Thomas. "I Am the Law." *American Heritage* 20 (June 1969): 32–48.

———. *Mysteries of My Father.* Hoboken, NJ: Wiley, 2005.

Flink, James J. *The Car Culture.* Cambridge: MIT Press, 1975.

Fogelston, Robert M. *Downtown: Its Rise and Fall, 1890–1950.* New Haven: Yale University Press, 2001.

Gasparini, D. A., and Judith Wang. "Battery–Joralemon Street Tunnel." *Journal of Performance of Constructed Facilities* 20, no. 1 (February 1, 2006): 92–106.

Gillette, Halbert P. *Handbook of Cost Data for Contractors and Engineers*. 2nd ed. New York: McGraw-Hill, 1910; revised 1918.

Hart, Steven. *The Last Three Miles: Politics, Murder, and the Construction of America's First Superhighway*. New York: New York Press, 2007.

Hewett, B. H. M., and S. Johannesson. *Shield and Compressed Air Tunneling*. New York: McGraw-Hill, 1922.

A History of the City of Newark New Jersey, Embracing Practically Two and a Half Centuries, 1666–1913. Vol. 3. New York: Lewis, 1913.

Jackson, Robert W. *Rails across the Mississippi: A History of the St. Louis Bridge*. Urbana: University of Illinois Press, 2001.

Johnson, Nelson. *Boardwalk Empire: The Birth, High Times, and Corruption of Atlantic City*. Medford, NJ: Plexus, 2002.

Konvitz, Joseph W. "William J. Wilgus and Engineering Projects to Improve the Port of New York, 1900–1930." *Technology and Culture* 30 (1989): 398–425.

Leinwand, Gerald. *Mackerels in the Moonlight: Four Corrupt American Mayors*. Jefferson, NC: McFarland, 2004.

Mahoney, Joseph H. "Walter Evans Edge." In *The Governors of New Jersey, 1664–1974: Biographical Essays*, ed. Paul A. Stellhorn and Michael J. Birkner. Trenton, NJ: New Jersey Historical Commission, 1982.

McCullough, David. *The Path between the Seas: The Creation of the Panama Canal, 1870–1914*. New York: Touchstone, 1977.

McKean, Dayton David. *The Boss: The Hague Machine in Action*. Boston: Houghton Mifflin, 1940.

Miller, Edmund Walter. *From Canoe to Tunnel: A Sketch of the History of Transportation between Jersey City and New York, 1661–1909: A Souvenir of Tunnel Day, July 19, 1909*. Jersey City, NJ: A. J. Doan, 1909.

Muirheid, Walter Gregory. *Jersey City of Today: Its History, People, Trades, Commerce, Institutions and Industries*. 1910. Reprint, Englewood, NJ: Bergen, 1996.

Murray, Thomas Hamilton. *Booklet of Information Regarding the American-Irish Historical Society*. Boston and New York: American-Irish Historical Society, 1905.

National Board of Fire Underwriters. *The Holland Tunnel Chemical Fire: New Jersey–New York, May 13, 1949*. New York: National Board of Fire Underwriters, 1949.

New Jersey Interstate Bridge and Tunnel Commission. *Report of the New Jersey Interstate Bridge and Tunnel Commission to the Senate and General Assembly of New Jersey, 1921*. Trenton, NJ: MacCrellish and Quigley, 1921.

———. *Report of the New Jersey Interstate Bridge and Tunnel Commission to the Senate and General Assembly of the State of New Jersey, 1922*. Union Hill, NJ: Hudson, 1922.

———. *Report of the New Jersey Interstate Bridge and Tunnel Commission to the Senate and General Assembly of the State of New Jersey, 1924*. Trenton, NJ: MacCrellish and Quigley, 1924.

———. *Report of the New Jersey Interstate Bridge and Tunnel Commission to the*

Senate and General Assembly of the State of New Jersey, 1926. Trenton, NJ: Mac-Crellish and Quigley, 1926.

New York, New Jersey Port and Harbor Development Commission. *Joint Report with Comprehensive Plan and Recommendations.* Albany, NY: J. B. Lyon, 1920.

New York State Bridge and Tunnel Commission. *Eighth Report of the New York State Bridge and Tunnel Commission to the Legislature of the State of New York, 1918.* Albany, NY: J. B. Lyon, 1918.

———. *Report of the New York State Bridge and Tunnel Commission to the Governor and Legislature of the State of New York, 1920.* Albany, NY: J. B. Lyon, 1920.

———. "Statement of Total Traffic, Operation of the Holland Tunnel from November 13, 1927, to January 31st, 1929." Science, Industry and Business Library, New York Public Library.

———. "Statement of Total Traffic, Operation of the Holland Tunnel, 1930." Reproduced from typewritten copy. Science, Industry and Business Library, New York Public Library.

Public Service Corporation of New Jersey. *Report to the Executive Committee of the Public Service Corporation of New Jersey on the Proposed Vehicular Tunnel between the Cities of Jersey City and New York by the Special Committee Appointed to Investigate the Subject.* March 1917.

Rae, John B. *The Road and the Car in American Life.* Cambridge: MIT Press, 1971.

Rose, Mark H., Bruce E. Seely, and Paul F. Barrett. *The Best Transportation System in the World: Railroads, Trucks, Airlines, and American Public Policy in the Twentieth Century.* Columbus: Ohio State University Press, 2006.

Scannell's New Jersey's First Citizens: Biographies and Portraits of the Notable Living Men and Women of New Jersey with Informing Glimpses into the State's History and Affairs. Vol. 1. Paterson, NJ: J. J. Scannell, 1917.

Seely, Bruce E. *Building the American Highway System: Engineers as Policy Makers.* Philadelphia: Temple University Press, 1987.

Shockley, Jay. *Gansevoort Market Historic District Designation Report.* Edited by Mary Beth Betts. New York: New York City Landmarks Preservation Commission, 2003.

Smith, Al. *Up to Now: An Autobiography.* New York: Viking, 1920.

Waddell, J. A. L. "Bridge versus Tunnel for the Proposed Hudson River Crossing at New York City." *Transactions of the American Society of Civil Engineers* 84, no. 1 (1921): 570–574.

Wilgus, William J. *Proposed New Railway System for the Transportation and Distribution of Freight by Improved Methods in the City and Port of New York.* New York: Amsterdam, 1908.

Winfield, Charles H.. *History of the County of Hudson, New Jersey, from Its Earliest Settlement to the Present Time.* New York: Kennard and Hay, 1874.

Wolfert, Ira. "Huge Tunnel Built Like a Double-Barreled Shotgun." *Popular Science,* October 1957.

Index

About the Author

Robert W. Jackson is an urban planner and historian with an interest in the built urban environment. As a historian for the Historic American Engineering Record, National Park Service, he has documented historic bridges and highways in Texas, Iowa, and Pennsylvania.